Landmarks

Dr. Daniel C. Drucker, President ASME, presenting the Society's first Landmark plaque to Lt. Gen. Stanley R. Larsen, U.S. Army (retired), who accepted it on behalf of the Mayor and the City of San Francisco on November 30, 1973.

National Historic Mechanical Engineering Landmarks

Richard S. Hartenberg, Editor

Sponsored by

The History and Heritage Committee

THE AMERICAN SOCIETY OF MECHANICAL ENGINEERS

United Engineering Center • 345 East 47th Street • New York, N.Y. 10017

Library of Congress Catalog Card Number 78-64993

About This Book . . . .

The History and Heritage Committee of The American Society of Mechanical Engineers was formed in 1971 by action of the Council. The Committee's purpose is to examine the mechanical engineering attainments of the past and, by recording the especially meritorious efforts, call attention to a rich heritage. ASME has identified these laudable accomplishments as National Historic Mechanical Engineering Landmarks to help define the patrimony from the engineering fathers. On occasion there has been a confluence of interests, as when the American Society of Civil Engineers or the Institute of Electrical and Electronics Engineers would find their Landmark qualifications as well. This volume is the registry of the first group of Landmarks, the twenty-eight recognized with appropriate ceremonies and distinguished by bronze plaques, between November 1973 and December 1977, in seventeen states. The Committee Chairmen who guided the Landmark designations were Eugene S. Ferguson and Donald E. Marlowe.

America's technological history is short when compared with that of Europe and lands more ancient still. Our history starts with the great migrations from England in the 1630 s. It manifested itself first in the integrated iron-works called Hammersmith established in 1647 at Saugus, just outside Boston. Since then, much has happened to give America a heritage of consequence. It is one that we want to cherish and keep alive in the minds and hearts of our colleagues of today, our successors of tomorrow, and present to the general public as well.

We have then a collection of sketches dealing with significant mechanical happenings of the past, things and events that provide a heritage we can enjoy, admire, and learn from. We discover how some familiar things began, but we also detect situations unfamiliar now because of the progress made over the several centuries. So it's a mixed bag of adventures much more important in their time than the rude states-of-the-arts might suggest to our sleek chrome-plated and plastic world. We can only marvel at the old time achievements on realizing they were accomplished without the handbooks, theoretical foundations, and encyclopaedic knowledge at our disposal today; as a matter of fact, our funds of knowledge are the outgrowths of the seat-of-the-pants engineering we look back to.

The Landmarks had their origins in the History and Heritage Committees of the Local Sections the country over. It is in these Committees that the "discoveries" of what might have Landmark status are made, and the evaluation studies pursued, to effect the nomination of the potential Landmark to the National Committee. This Landmark program is a grass roots affair, for the Local Sections not only set out the Landmarks, but arrange for a ceremony, and a brochure telling the story of the occasion.

The Landmark sketches that follow are in the main drawn from the brochures supplied by the local History and Heritage Committees. Countless individuals have contributed original data and formed it into stories that for the purposes of this compilation needed some reshaping. The Editor is of necessity a plagiarist, lifting much of the final copy from the works of many, mutilating and truncating them in the process. I happily acknowledge my dependence and debt upon nearly two-score names, and knowing that the list is bound to be incomplete, ask the pardon of those overlooked through inadvertence. Warmest thanks are rendered to the following: John D. Alden, Joyce Baida, F. W. Beichley, Don Belanger, Walter A. Biddle, Francis J. Blesso, Tom Bostwick, T. E. Bowman, Otto Bull, John G. Burke, Colin Carmichael, C. Glenn Crawford, Kenneth R. Daniel, R. Carson Dalzell, David Dehring, A. Dil Pare, Fred Ellis, J. J. Ermenc, Louis T. Frantz, Martin R. Frick, H. B. Garrett, Carron Garvin-Donohue, George F. Gayer, James C. Glascock, Hudson C. Haile, Elmer M. Johnston, B. A. Kilgore, Gary Kulik, Forrest E. Lange, Patrick M. Malone, J. L. Mangan, James J. Matera, Otto Mayr, P. F. McCawley, Reid McNulley, Robert F. Metcalf, Conrad Milster, Les E. Muenster, Carey Murphy, Gus Nick, Francis O'Keefe, Robert E. Perreault, John W. Prud'homme, J. D. Quinn, Linwood C. Robinson, George K. Saurwein, Bernadette Savard, Peter Serratore, C. P. Shaw, E.F.C. Somerscales, R. C. Spencer, Jack R. Stokvis, E. Vandergrift, Jr., R. M. Vogel, F. W. Von Flue, Laurence E. Webber, Chris White, Walter C. Woodman. There is a very special indebtedness to Maurice Jones, ASME's former Director of Public Relations, whose enthusiasm overcame many crises.

By way of introduction to the Landmarks there is "A Sense of the Past," an essay written in 1974 about ASME's many concerns with matters of historical relevance. The author is Eugene S. Ferguson, Professor of History at the University of Delaware, Curator of Technology of the Hagley Museum, and first Chairman of the History and Heritage Committee. The late Colonel Clarence E. Davies, long-time Secretary of the Society, always a moving force in history and heritage issues, and possessor of an extraordinary memory, contributed bits of personal experiences to Ferguson's treatise. The roster of ASME's involvements is seen to include a wide range of subjects of which the activities of the History and Heritage Committee and its Landmark program are firmly rooted in a complex of historically related topics.

Richard S. Hartenberg

The Technological Institute
Northwestern University
Evanston, Illinois

Photo by Bruce Ross

About the Editor

Richard S. Hartenberg

- Fellow ASME
- Fellow Institution of Mechanical Engineers
- Verein Deutscher Ingenieure
- Professor Emeritus of Mechanical Engineering, Northwestern University

Dr. Hartenberg began his academic career at the University of Wisconsin. His special interests are in Kinematics and the History of Technology. He now serves as advisory editor on two international quarterlies, *Mechanism and Machine Theory* and *Technology and Culture.*

A long activity in ASME affairs includes present membership in the Design Engineering Division's Senate and the National History and Heritage Committee.

Contents

About This Book v
Richard S. Hartenberg

About the Editor ix

A Sense of the Past xv
Eugene S. Ferguson

National Landmarks

Twenty-eight National Landmarks, designated during the period of 1973-1977, are listed in the order of recognition.

Ferries and Cliff House Cable Railway — 1887 3
San Francisco, California

Leavitt Pumping Engine — 1894 9
Brookline, Massachusetts

A. B. Wood Low-Head High-Volume Screw Pump — 1915 13
New Orleans, Louisiana

Portsmouth-Kittery Naval Shipbuilding Activity — 1774 17
Portsmouth, New Hampshire

102 Inch Boyden Hydraulic Turbines — 1871/1873 23
Cohoes, New York

5000 kW Vertical Curtis Steam-Turbine Generator — 1903 29
Schenectady, New York

Saugus Iron Works — 1647 33
Saugus, Massachusetts

Pioneer Oil Refinery — 1876 39
Newhall, California

Chesapeake and Delaware Canal Pumping Machinery — 1851 43
Chesapeake City, Maryland

Reciprocating Steam Engines USS TEXAS — 1912 47
Houston, Texas

Childs-Irving Hydroelectric Project — 1909/1916 51
Irving, Arizona

Hanford B-Reactor — 1944 55
Hanford, Washington

Underground Air Conditioning—Magma Copper Mine — 1937 59
Superior, Arizona

Pike's Peak Cog Railway — 1891 63
Colorado Springs, Colorado

Edgar Steam-Electric Station — 1925 69
Weymouth, Massachusetts

Mount Washington Cog Railway (Joint ASME/ASCE Landmark) — 1869 73
Mount Washington, New Hampshire

Folsom Powerhouse No. 1 — 1895 79
Folsom, California

Crawler Transporters of Launch Complex 39 — 1965 83
J.F.K. Space Center, Florida

Fairmount Water Works (Joint ASME/ASCE Landmark) — 1815/1911 87
Philadelphia, Pennsylvania

Vertical Reciprocating Steam Engines USS OLYMPIA – 1891 95
Philadelphia, Pennsylvania

Five-Ton "Pit-Cast" Jib Crane – 1905 99
Birmingham, Alabama

State Line Generating Unit No. 1 – 1929 105
Hammond, Indiana

Pratt Institute Power Generating Plant – 1887/1900 109
Brooklyn, New York

The Monongahela and Duquesne Inclines – 1870 and 1873 115
Pittsburgh, Pennsylvania

Great Falls Raceway and Power System (Joint ASME/ASCE Landmark) – 1794 121
Paterson, New Jersey

Vulcan Street Power Plant (Joint ASME/ASCE/IEEE Landmark) – 1882 125
Appleton, Wisconsin

Wilkinson Mill – 1811 131
Pawtucket, Rhode Island

Local Landmarks, 1973 – 1977

Paige Compositor – 1887 141
Hartford, Connecticut

Reynolds-Corliss Engine and Pump – 1917 145
Jacksonville, Florida

A Sense of the Past

Historical Publications of The American Society of Mechanical Engineers

Eugene S. Ferguson, Member ASME

Summary

The American Society of Mechanical Engineers has both a tradition and a solid record of encouraging and supporting historical publications, particularly in the field of biography but also in the history of technical achievements. The numerous books, articles, and commemorative brochures that have been published under Society auspices or with the encouragement of the Society provide a great deal of historical information, much of which would not be otherwise available. These publications make collectively a substantial and tangible witness to the sense of the past that has inspired and in some measure informed a surprisingly large number of prominent individuals who are or have been members of the Society.

A history of the Society was planned in 1908 and published in 1915 in an attractive and durable volume. The celebration in 1930 of the Society's fiftieth anniversary was marked by a special number of *Mechanical Engineering.* A series of book-length biographies of individuals was commenced in 1912 and continued into the nineteen-fifties; in all, sixteen volumes were published. A valuable series of informative obituary notices, often the only dependable surviving source of information regarding an individual engineer, was started very early in the Society's life and continued systematically until 1949; occasional extended notices have since appeared in *Mechanical Engineering.*

A number of significant historical articles, written by men who took part in or knew at first hand the development they described, were published in the early *Transactions* and, after its commencement in 1906, the *Proceedings* (later renamed *Journal* and, in 1919, *Mechanical Engineering*).

Since 1971, a program to encourage appreciation of the history and heritage of mechanical engineering, including the supervision of a National Historic Mechanical Engineering Landmark program, has been undertaken by the Society's History and Heritage Committee. That Committee sees the past as a resource that can be husbanded in various ways. Technical events within the purview of ASME can furnish information of permanent importance and significance. Surviving machines, drawings, photographs, and other objective evidence of accomplishment can convey to future generations some sense of our aspirations and abilities. To the present generation the past offers insights and wisdom that our predecessors spent their lives in obtaining.

A current statement of the Society's sense of the past was made in 1974 before a joint committee of the United States Congress by Donald E. Marlowe, Past President of ASME and currently (1974) Chairman of the History and Heritage Committee.

> For the first half of this century, it has seemed that technological advance was inevitable and that it was always good and that it was limitless. We are now aware that none of these statements are true. But as we contemplate an era in which limited human and material resources will have to be deployed with great care if maximum service to the people is to be secured, we find ourselves desperately seeking guidance from the past. To be very blunt, we do not know what policies will influence the flow of innovation, how to diffuse the benefits of technology into depressed or primitive areas, how to adjust to true technological breakthroughs, how to distinguish between net and gross in technological progress, what the necessary relationship between science and technology is, and many other equally important questions. Only in the last few years have we begun to devote to the history of technology the attention it deserves, for this history, far more than that of kings or ministers, can help us to understand the social impact of technology, and perhaps assist us in setting our course for the future [1].

Society History

A history of the ASME was projected in 1908, when a committee made up of John E. Sweet, a founder of the Society, Henry Harrison Suplee, editor of *Cassier's Magazine,* and Charles W. Hunt published in ASME *Proceedings* an account of the Society's founding (in 1880), which was in effect the first installment of an ambitious history of the Society. Suplee gathered more material for the history; eventually Frederick R. Hutton, Secretary of the Society from 1883 to 1906 and its president in 1906, took over the writing and produced in 1915 an attractive and informative volume, copiously illustrated. Clear, full-page portraits of each of the first thirty-four presidents, three treasurers, and two secretaries were included. The several meeting rooms and buildings occupied by the Society were illustrated, including the Engineering Societies' Building on 39th Street, dedicated in 1907 [2].

A concern with historical objects and memorabilia as well as a narrative record was evident in the *History.* Many portraits, drawings, and other objects are mentioned and their locations occasionally specified in offices, library, and other public spaces. Some of the objects were probably misplaced or lost when the Society moved to its present quarters, but an inventory of historical portraits and other objects might recapture a number of things now unknown or forgotten [3].

The historical development of the Boiler Code, one of ASME's proudest achievements, was recorded by Arthur M. Greene, Jr., a member of the Boiler and Pressure Vessel Committee from 1915 to 1943, published serially in *Mechanical Engineering,* and published separately as a substantial and attractive volume [4][a]. A historical summary of the Diesel and Gas Engine Power Division was prepared recently by T. M. Robie for the fiftieth anniversary of the Division's founding. Robie's bibliography lists two or three earlier historical papers presented at division meetings [5]. As yet unpublished, Ralph A. Sherman's "Fuels Division of the ASME 1920–1972" is an informative and perceptive historical essay [6][b].

Anniversary Publications

A Fortieth Anniversary Meeting of ASME, held in New York in 1920, was addressed by Henry Towne, oldest past president (1889) of the Society and others, including J. Herbert Case, of the New York Federal Reserve Bank, and, as a conciliatory gesture, Samuel Gompers, perennial president of the American Federation of Labor. The proceedings were reported in *Mechanical Engineering* [7].

a) Clarence E. Davies, ASME's long-time Secretary, added some remarks: "The ASME Council was concerned about the headstrong attitude of John A. Stevens, Chairman of the Boiler Code Committee in 1915, and put Arthur M. Greene, a Council Member (Professor of ME at Rensselaer) on the Boiler Code Committee to keep Stevens in line. Greene was later Dean of Engineering at Princeton and his successor as Dean, Kenneth Condit, asked me to find something to keep Greene busy because his active mind kept him on the move and he was getting into Condit's hair. I suggested that he write the Boiler Code history. The Boiler and Pressure Vessel Committee gladly assented and Greene went to work. Chapters of the history were carried in *Mechanical Engineering* in 1952 and the complete book appeared shortly thereafter."

b) Clarence Davies again: "The mention of Robie's history of the Diesel, and Gas Engine Power Division points to the "History of the Fuels Division" by Ralph A. Sherman, published in draft in 1972. This document by an expert reports the meetings of the Fuel Division but also gives an interesting account of the changes in combustion of coal. From about 1907 until 1915, the ASME had a Gas Power Section and a list of those affiliated with the Section appeared in the Annual ASME Membership Lists. I recall that Dr. Charles E. Lucke of Columbia wrote a history of this operation. Hutton, in the "History," mentions the Gas Power Section on page 294 in his discussion of professional sections. I am making a serious search for the Lucke paper."

A Carnot Centenary Commemoration was held in New York in December, 1924, and reported in *Mechanical Engineering.* Sadi Carnot's seminal book *On the Motive Power of Heat,* published in 1824, had been translated into English by Robert Thurston, first president of ASME, and published by him in 1890, with a second edition in 1897. Thurston's translation was republished by the Society in 1939 with a further edition in 1959 [8].

In 1927, the Council of ASME announced that "plans are already in hand for the Spring Meeting and the Semicentennial" of the Society, to be held in April of 1930, "commemorating the first meeting fifty years ago, April 7, 1880, held at Stevens Institute of Technology" [9].

A special Fiftieth Anniversary issue of *Mechanical Engineering* was published in April, 1930. The first part of the issue, entitled "To Perpetuate Great Names," included over a hundred portraits of presidents, honorary members, and others. The second part of the issue contained "a group of reports presenting the history of technical development since 1880"; individual reports dealt with aspects of power, manufacturing, transportation, applied mechanics, and engineering education.

The May issue of *Mechanical Engineering* reported the Anniversary Celebration, which started in New York and then moved to the National Council Chamber of the U.S. Chamber of Commerce, in Washington. A pageant, written by a Yale professor and given by students of Stevens Institute of Technology, celebrated "the increasing triumphs of the mechanical engineer." Among the members of the cast were Intelligence, Imagination, Conversion, and Finance; holding central sway was Mature Control. In Washington, the representatives of foreign technical societies gave prepared papers on the "social and economic effects of engineering upon the lives of people" during the preceding half-century in their respective countries. President Herbert Hoover, a member of

ASME, addressed the Anniversary Dinner in the Mayflower Hotel, and a reception was held in the White House [10] [c].

A hard-cover volume, published as a result of the Society's Fiftieth Anniversary, was edited for the Anniversary Committee by the well-known historian Charles A. Beard. In the Council Report of 1929 the plans for the volume were explained:

> Much thought has been given to publications commemorative of the Society's Fiftieth Anniversary to be celebrated in April, 1930. The Committee has received the enthusiastic interest and cooperation of Dr. Charles A. Beard in the preparation of a book which will be a symposium published by Longmans, Green, and Company under the title *Toward Civilization.* The contributors to this symposium are eminent engineers and scientists, and they will attempt to answer from the point of view of the technologist the many indictments which have been leveled at the machine civilization of today by philosophers and humanists [11].

The particular provocation for this project was a volume edited in 1928 by Beard and published under the title *Whither Mankind.* That book, consisting of essays by historians, philosophers, and others, contained "a challenge" which was taken seriously by Clarence Davies, then Assistant Secretary of the Society. Davies persuaded both editor and publishers to make the book and helped line up the contributors [12] [d].

In 1930, *Toward Civilization* took its place as a companion volume to *Whither Mankind.* Although neither collection of essays contained anything that in 1930 would have surprised or instructed an informed reader, the occasion served to elicit statements of faith of several prominent mechanical engineers, such as Ralph Flanders, Lillian Gilbreth, and Elmer Sperry, in the efficacy of a technical approach to human problems [13].

Another commemorative publication designed to answer critics of engineers was the brochure that in 1937 recorded the proceedings of the "forum presenting the career and achievements of George Westinghouse on the 90th anniversary of his birth." Conrad Lauer, a past president of ASME, had written in 1934:

> The engineer is on the defensive in justifying his work and establishing himself as a constructive element in civilization and a driving force in providing for the well-being of mankind.
>
> The American Society of Mechanical Engineers should dramatize the contributions of the outstanding leaders in engineering invention, design, construction, and application [14].

---

c) Having been at the anniversary celebration, Clarence Davies recalled the following: "The 50th Anniversary meetings started on Saturday with a luncheon at McGraw-Hill to recognize the contribution of *The American Machinist* on the founding of the Society. It then moved to Hoboken where the pageant was given in the Great Hall at Stevens Institute where the first meeting of the Society was held. Then back to New York for a banquet Saturday evening at the Roosevelt. Sunday there was a special service at St. John the Divine and a special train to Washington. Elmer Sperry was stricken on the train and died a few days later. The representatives of foreign societies who gave the papers at Washington were granted special 50th Anniversary Medals and were given honoraria of $1000 each. At the ceremony when the medals were presented, each foreign representative was escorted to the dais by the ambassador from his country. At the banquet, there was no head table, only a platform. There were [16] honor tables at each of which the foreign representative and his ambassador were the honor guests. Also at the banquet the Hoover Medal was inaugurated by the presentation of the first Hoover Medal to Mr. Hoover."

d) Clarence Davies had this to say: "The challenge to *Whither Mankind* was taken seriously by 'me,' who aroused the interest of the Publications Committee, particularly William A. Shoudy who worked assiduously with me to interest Charles Beard and to secure the contributors to *Toward Civilization.*"

The Westinghouse commemorative brochure[e] contained testimonials of prominent men, some of whom were former associates of Westinghouse.

Biographical Series

The first biographical volume to be published was the autobiography of John Fritz, a patriarch of the iron and steel industry, who at ninety, in 1912, reluctantly agreed to see in print what he had intended as a posthumous testament [15].

e) A fill-in from Clarence Davies: "The Westinghouse commemorative publication was prepared for the specific purpose of producing a document for use in the Society campaign to get Westinghouse into the Hall of Fame for Great Americans at New York University. Charles F. Scott at Yale was the public leader of this effort. Scott had worked for Westinghouse. He also knew Andrew Carnegie and with Calvin Rice, they were the principal instruments in getting $1,500,000 from Carnegie for the old Engineering Societies building on 39th Street and the building for the Engineers Club on 40th Street. Herman Westinghouse, brother of George, presented a bust of George, by Daniel Chester French, to the Society with the stipulation that when George was eligible for the Hall of Fame, the Society would bend its efforts to get George Westinghouse elected to the Hall of Fame. Westinghouse was eligible in 1940 and under Scott's leadership we conducted a campaign which resulted in 38 votes. The 1945 and 1950 campaigns resulted in 44 and 34 votes. The 1955 campaign was mounted as the final effort and arrangements were made to interview all of the 100 Electors. 62 favorable votes resulted and on December 1, 1957 the bust of George Westinghouse was placed in the Hall of Fame. The bust of French held by ASME was not used, however, because by 1957 French had his alloted number of busts in the Hall of Fame and the French bust, while at ASME, had been exposed to public view. This violated the rule that the bust for the Hall of Fame must not be exposed to public view until it is unveiled at the installation ceremony. You can see the French bust on the 6th floor of the United Engineering Center. It is interesting to note that Westinghouse was admitted to the Hall of Fame as an inventor, not an engineer. (There is only one engineer in the Hall of Fame, James Buchanan Eads). At the installation ceremony a bust by Quattrochi was presented by William F. Ryan, President of ASME; Herbert Hoover made the address."

The evident worth of the book as the record of an important and widely revered engineer led in time to an active program of publishing book-length biographies. The 1916 Council Report promised that "a Biographies Committee is to be appointed to supervise the preparation and editing of biographies of noted engineers [16]. Ad hoc committees supervised the biographies of George Westinghouse (1921), Frederick Winslow Taylor (1923), and John A. Brashear (1924) [17]; by 1925 a Biography Advisory Committee had been formed; over the next thirty years, with the practical assistance of Secretaries Calvin Rice and Clarence Davies, the Committee saw to the publication of an impressive series of biographies and autobiographies. In Rice's review, in 1930, of the Society's history, he noted the publication to date of eight biographies. The aim of the Biography Advisory Committee, he wrote, was "to add wisely to this brief list as years go on, believing that such biography is not only an important contribution to engineering literature, but that it is an unsurpassed source of inspiration for younger men [18]." One remarkable aspect of the biography program is the number of prominent and busy men who thought it important. The committee, during the thirty-odd years it lasted, was served by such men as W. F. M. Goss, dean of Purdue's engineering school, Ralph Flanders, for many years a United States Senator from Vermont, Fred R. Low, George Orrok, and Roy V. Wright.

Financing the publication of biographies was a perennial problem. Some of the volumes were paid for by friends or associates of the subject; several of them were issued in both a subscription edition and a trade edition, the latter presumably at the risk of the commercial publisher involved [19]. It is perhaps significant that men who so often, in their professional pursuits, raised the question "Will it pay?" would invest so much of their energy in a program that was unlikely ever to break even.

A list of the biographies published under ASME auspices or with its encouragement appears below, in Appendix A.

Necrology

The series of book-length biographies is clearly an outstanding contribution to our historical understanding, yet the Society's necrology, another biographical undertaking which was sedulously pursued during nearly the first sixty-five years of the Society's existence, provides now an irreplaceable record of individual mechanical engineers. Informative obituary notices appeared in the *Transactions* through 1926, in the *Record and Index* during the three years it was published (1927 through 1929), and again in the *Transactions* through 1949. Obituaries will be found in the earlier *Proceedings* and *Journal,* and a number of biographical sketches have been published in *Mechanical Engineering,* but as the Society grew and the Necropolis became overcrowded the burden of recognizing individual careers was apparently no longer a tolerable one.

A volume of lasting value to historians of technology and other scholars could be readily assembled by simply reproducing all the obituary notices that have been published in the Society's periodicals.

Reminiscences

Extended discussions of papers in the early *Transactions* add depth and understanding to the matters treated in the papers. Many personal recollections were recorded in this way. A classic description by Alfred C. Hobbs of his successful opening in1851 in London, of the famous Bramah lock, appears in the discussion of a paper on American machinery at international exhibitions [20]. The incident apparently encouraged Hobbs, a lively storyteller, to prepare in the following year a paper on "Locks and their Failings" [21]. Robert Allison, on the other hand, contributed in 1895 a paper on "the difficulties and trials of the old-time machinists, of which the writer was one." Although the account is valuable in the permanent record and much discussion followed the talk, it was probably merely tedious to younger listeners [22].

Some further recollections come from Clarence Davies:

• John Clinton Parker of Philadelphia was a critic of the ASME Boiler Code starting in 1915 before the Code was published and continuing until his death in 1946. In 1934 he started action in the Supreme Court of New York County, first requiring a review of all Society activities and later bringing suit against 47 members of the Council for the return to the Society treasury of $260,000 spent in developing the Engineering Index which he claimed was not provided for in the Society's charter. He won the case in the lower court but the decision was overturned by the Appellate Division and the Court of Appeals in 1939. The case cost the Society $30,000 in legal fees and the only tangible result from it was the ability to quote the following as a footnote to the ASME Certificate of Incorporation: "That in promoting the arts and sciences connected with mechanical construction for scientific purposes, The American Society of Mechanical Engineers may employ any reasonable means to accomplish these ends and it is not restricted to the specific means stated in its charter."

The "Parker Case" took five years of my time as I started in as the new Secretary in 1934 and was full of business recovering from the depression—keeping the society in the black, organizing the Engineers Council for Professional Development, among many other things. The "Parker Case" may be worth a chapter in a future Society History.

• The Society's 75th Anniversary in 1955 was a national year-long celebration, each of eight meetings during the year being exciting occasions also worth a chapter in a future Society History.

• The United Engineering Center at 345 East 47th Street started in earnest on February 12, 1957 when I loaded the United Engineering Trustees into rental cars and visited all the available sites in New York City. At noon the Real Estate Committee met and voted to recommend the present site to the Trustees. At lunch, the Board of UET voted to approve the site. Work was started the next day to buy the site which was finally secured in August. ASME moved in Labor Day, 1961. I was Executive Director at the new Engineering Center from 1957 to 1961.

History Articles

Significant historical articles have appeared from time to time in the *Transactions* and in *Mechanical Engineering.* More than fifty articles are listed under "History" rubrics in the ASME *Seventy-Seven Year Index 1880–1956* (1957), which are reproduced in Appendix B. It should be noted that these lists are not exhaustive since many history articles are listed under appropriate subject headings rather than "History."

Of particular note are the articles which supply fresh, often unique, information. Articles on rolling mills by C. H. Morgan (1901), on high-speed engines by Charles T. Porter (1901), and on hydraulic turbines by Edward Dean Adams (1930), are examples of those written by men whose own careers helped make the history they recorded. The extended article on the work of Erasmus D. Leavitt, published in the *Transactions* shortly after his death, is a permanently valuable contribution [23]. In 1886, George H. Babcock, one of the founders of Babcock & Wilcox Company, published the article "Substitutes for Steam," a fresh and comprehensive historical review of attempts to use naptha, ammonia, and other fluids in a vapor cycle in order to improve its economy [24].

Three historical articles on turbines were sponsored at the Annual Meeting in 1936 by the "Joint Division on Engineering History" and subsequently published in *Mechanical Engineering.* The papers treated the historical development of the turbines of, respectively, Westinghouse, General Electric, and Allis-Chalmers [25].

A number of pedestrian papers on the engineer's historic contributions to civilization have also appeared in the Society's publications. It is difficult for an editor to say "no" to a prominent member of the Society whose critical faculties do not extend to his own version of history. Unfortunately, such "schoolboy" history, which is part of a culture's mythology (for example, Watt invented the steam engine, Fulton the steamboat; engineers were ignorant and simple-minded until they learned mathematical methods), repels intelligent engineering students and confirms their notion that history can have nothing of value to say to them [26].

Engineering History Committees

A persistent interest in the study of engineering history will be found in the Society's publications. A totally unexplained "Historical Guild" appeared as one of the special administrative committees in a Council Report of 1924, but it did not outlast the year [27].

In 1934, two dozen ASME members proposed to form within the Founder Societies (ASCE, AIME, ASME, AIEE)

> "a body to encourage the study of the history of early American engineers and engineering and technology in the United States. The plan is to bring together engineers in all branches of the profession who are interested in engineering history, who are engaged in historical research, or who are interested in encouraging the recording of historic developments for the benefit of those who will be engaged in historical research at a later date . . . . "

A charter membership of at least two hundred, with annual dues not to exceed four dollars, was the goal [28]. The proposal, coming as it did deep in the Great Depression, did not elicit the hoped-for response.[f] The "Joint Division on Engineering History", mentioned above as sponsors of the 1936 session on steam turbine history, was no doubt an outgrowth of this proposal, but the relationships are not clear.

f) Clarence Davies on the way it was: The "depression" was a rugged time for ASME from about 1932 through 1935. We were short of cash and paid our rent with notes. We sold bonds to the members. Rice and I took 30 per cent cuts in salaries. All staff people suffered reduced pay and many of the staff were separated. This cannot be read in *Mechanical Engineering.*

Museums Committees

Early in 1913, the ASME "Committee on National Museum" reported that a German commission of five members, "with as many engineer assistants," had been sent to the United States by the Deutsches Museum, Munich, to "look for more contributions of machines, models, drawings, etc., to enable them to add an American section to each of their main divisions and to encourage American engineers to begin a similar movement." Attempting to arouse interest in a "great National Museum of the Industrial Arts," the Committee saw its immediate aim to "preserve the landmarks of our journey to a higher civilization and thus write in enduring materials of our own fashioning, the true history of the American people" [29].

Ten years later, an editorial in *Mechanical Engineering*, entitled "Engineering Museum Requires Interest and Aid of Entire Profession," explained how the establishment of a National Museum of Engineering and Industry was now in the hands of a committee representing all the national engineering societies [30]. In 1924, a pretentious brochure, issued from the Engineering Societies Building in New York, outlined plans for the national museum as an extension of the Smithsonian Institution in Washington, with branch museums in several cities. The first two branches were to be in Chicago and New York. In Chicago, Julius Rosenwald had promised a million dollars, which was eventually used in development of the Chicago Museum of Science and Industry; in New York, a bequest of ASME past-president Henry R. Towne was to be used to establish a "Museum of the Peaceful Arts," which eventually opened as the short-lived Museum of Science and Industry [31].

History and Heritage Committee

Formed in 1971, the present History and Heritage Committee spent its early years in an attempt to interest ASME local sections in learning, by conducting surveys and making inventories, of historically significant and interesting local survivals of particular pertinence to the practice and development of mechanical engineering. An extensive listing of early mill-sites and related mechanical engineering works and artifacts in Northern New England was completed in 1973 by the Northern New England Section under the leadership and superintendence of Forrest E. Lange, Chairman of the Section's History and Heritage Committee. The resulting *Historic Engineering Record* has been published in an attractive brochure [32].

A National Historic Mechanical Engineering Landmark program has been undertaken by the national committee in support of local History and Heritage Committees; initially, the response has been enthusiastic. The first national landmark is the San Francisco Cliff House and Ferries Cable Railway (1887), the second the steam-driven water-pumping engine (1894) designed by Erasmus D. Leavitt and located in the Chestnut Hill Pumping Station in suburban Boston. The third national landmark is the Wood high-volume screw pump (1912) of the New Orleans Sewerage and Water Board. In each of these dedication ceremonies Daniel Drucker, President of ASME, presented bronze plaques to the appropriate authorities; each ceremony gave rise to a publication [33].

Conclusion

The published historical work reviewed here provides a great deal of historical information that would otherwise be unavailable. It represents, moreover, "A Sense of the Past" on the part of a significant number of ASME members, many of them leaders of the Society.

As the present review suggests, the History and Heritage Committee breaks no new ground as it encourages the publication of historical information. It can, however, help to raise the quality of the history to a new level and increase greatly its interest and apparent relevance to many engineers. This will require the recognition of a discipline of history and a willingness to deal thoughtfully and responsibly with the questions of values and meaning. The discipline of history involves analysis different in kind from that of the engineer; its variables cannot be quantified without distortion or impoverishment; and attempts to standardize analytical operations are destined to yield absurdities. History deals, moreover, with meanings and moral values. Whenever meaning and values are involved, honest men of good will can differ, and varying interpretations will be made of the same event. Historical questions worth asking always have ambiguous answers, and the answers provide us not with conclusions as much as a basis for reflection and further discussion.

The training of engineers tends to suppress rather than encourage a sense of history, but it is clear that many engineers transcend their academic background. As Donald Marlowe has pointed out, the history an engineer learns gives him no sense of its ambiguity. "The skimpy historical education of engineers," he writes, "leaves them with a firm belief that (without qualification) George Washington *was* the father of his country; Eli Whitney *was* the father of mass production; Fulton *did* invent the steamboat. A little knowledge is, indeed, a very dangerous thing" [34].

While the History and Heritage Committee will clearly serve a useful purpose in planning and coordinating a National Historic Mechanical Engineering Landmark program, its lasting importance will probably depend chiefly upon its success in encouraging and developing, critical and relevant historical publications.

Reference Notes

[1] From typescript of statement made in April 1974, before the Subcommittee on the Department of Interior and Related Agencies.

[2] Hutton, Frederick Remsen, *A History of the American Society of Mechanical Engineers from 1880 to 1915* (New York: ASME, 1915) 355 pp., illus.

[3] Ibid, pp. 296-306. In 1959 or 1960, I collected for the National Museum of History and Technology the "hook tool," a long-handled, hand-held, lathe tool for turning metal, donated to the Society by John Fritz. A number of other objects that I asked about could not then be readily located.

[4] Arthur M. Greene, Jr., *History of the ASME Boiler Code* (New York, c. 1955), 145 pp.

[5] Robie, T.M., *A Fifty Year History of the Diesel and Gas Power Division . . . 1921-1971* (N.p., 1971). 21 pp.

[6] A typescript of Sherman's history is in the files of the ASME History and Heritage Committee. Sherman viewed in the light of an energy crisis a generation's activities in promoting unlimited use of irreplaceable resources. His essay should be published.

[7] *Mechanical Engineering,* XLII (Dec. 1920), 712-715.

[8] *Mechanical Engineering,* XLVII (Jan. 1925), 37-40; William F. Durand, *Robert Henry Thurston* (New York: ASME, 1929), p. 196; Clarence Davies, in a memorandum of October 2, 1973, to Rogers Finch, recalled that "The Carnot book was included because John Roy of Baltimore was interested in fine typography and he hand-set the book in type. We published the book as a gift token for the Society to present to important foreign engineers. The first edition, published in 1939, had a foreword by A. G. Christie, then ASME President. A second edition came out in 1959 when Walker Cisler was President, and he wrote an additional foreword."

[9] ASME, *Record and Index,* I (1927), 24.

[10] *Mechanical Engineering,* LII (May 1930), 509-528. It may be recalled that the ASCE Centennial Celebration in 1952, in Chicago's Museum of Science and Industry, also included an elaborate pageant. For a brief notice of advance planning for foreign participation (550 invitations to scientific and engineering societies, etc.), see ASME *Record and Index,* III (1929), 95-96.

[11] ASME, *Record and Index,* III (1929), 97.

[12] Beard, Charles A., ed., *Whither Mankind* (New York: Longmans, Green, 1928), 408 pp.; Clarence Davies to author, personal communication, Sept. 20, 1974.

[13] Beard, Charles A., ed., *Toward Civilization* (New York: Longmans, Green, 1930), 307 pp.

[14] American Society of Mechanical Engineers, *George Westinghouse Commemoration: A Forum Presenting the Career and Achievements of George Westinghouse on the 90th Anniversary of his Birth, December 1, 1936* (New York: ASME, 1937). 78 pp., illus. Quotation is from p. 5. Reprinted from *Mechanical Engineering,* LIX (Mar., Apr. 1937).

[15] Fritz, John, *Autobiography of John Fritz* (New York: ASME, 1912), p. vii.

[16] ASME *Transactions,* XXXVIII (1916), 20.

[17] The respective committees are named in the prefaces of the books. For full titles, see Appendix A.

[18] *Mechanical Engineering,* LII (April 1930), 266-267. Clarence Davies, in a memorandum of October 2, 1973 to Rogers Finch, wrote: "Calvin Rice, my predecessor, was deeply interested in the museum problem and he gave me great moral support on the biography program."

[19] The George Westinghouse biography was presumably financed by his brother, Henry Herman Westinghouse, who was "always in consultation" while the volume was being prepared; certainly he provided the money to produce the biography of Walter Craig Kerr (see prefatory statements in the respective volumes). An annual statement of Society finances in 1928 includes a line for the "Gleason Gift for Thurston biography." (*Record and Index,* II [1928], 95).

[20] Pickering, Thomas H., "American Machinery at International Exhibitions," ASME, *Transactions,* V (1884), 113-130; Hobb's story is related in pp. 123-130.

[21] Hobbs, Alfred C., "Locks and their Failings," ASME, *Transactions,* VI (1885), 233-235.

[22] Allison, Robert, "The Old and the New," ASME, *Transactions,* XVI (1895), 742-761. A series of lively, intensely interesting reminiscences was recorded by Charles T. Porter, of high-speed steam engine fame, and published serially in *Power* and *American Machinist.* Brought together in a volume in 1908 by John Wiley, the volume predated the Society's biography series. I hope that the book may be reprinted in the ASME series.

[23] Adams, Edward Dean, "Early Hydraulic-Turbine History," *Mechanical Engineering,* LII (Apr. 1930), 395-399; discussion, 719; correction, LIII (Jan. 1931), 67. F. W. Dean, "An Account of the Engineering Work of Erasmus Darwin Leavitt," ASME, *Transactions,* XXXIX (1918), 993-1036.

[24] Babcock, George H., "Substitutes for Steam," ASME, *Transactions,* VII (1886), 680-741.

[25] "The Steam Turbine in the United States," *Mechanical Engineering,* LVIII (Nov. 1936), 683-96; LIX (Jan., Apr. 1937), 71-82, 239-56.

[26] For example, a series of lectures on "Engineering's Part in the Development of Civilization," given in 1938 at the North Carolina State College, Raleigh, by Dugald C. Jackson, was published serially in *Mechanical Engineering,* LX (July-Dec. 1938), and reprinted in a duodecimo hardcover volume (New York, 1939). 114 pp.

[27] *ASME Transactions,* XLVI (1924), 10. The Historical Guild Committee included J. W. Lieb, chairman; D. S. Jacobus; R. Adm. R. S. Griffin; and J. W. Roe.

[28] The quotation is from a mimeographed sheet, dated May 31, 1934, kindly supplied by ASME Life Member William J. Ellender, who identified the twenty-four "proposers" as ASME member. See also I. N. Lipshitz, "The Study of Technical History; A Plea for an Organization in This Country Devoted to the Study of Engineering History," *Mechanical Engineering,* LVII (March 1935), 143-147. Lipshitz's paper was given at the 1934 Annual Meeting of ASME.

[29] "Report of Committee on National Museum," ASME, *Journal,* XXXV (Jan. 1913), 37-39.

[30] *Mechanical Engineering,* XLV (Feb. 1923), 138-139.

[31] *National Museum of Engineering and Industry* (New York, 1924), 24 pp. I am indebted to William J. Ellender, who loaned me a copy of this brochure. The Rosenwald and Towne gifts are mentioned on p. 21; the statement regarding the abortive New York Museum of Science and Industry is based upon my own recollection. In the aftermath of the Museum's failure, a number of unique early machines were lost and probably destroyed.

[32] Lange, Forrest F., ed., *Historic Engineering Record* (Portsmouth, N.H.: ASME, Northern New England Section, 1974), 28 pp. illus.

[33] Beichley, F. W., *Ferries and Cliff House Cable Railway, 1887* (San Francisco, Nov. 30, 1973), 12 pp. illus.; James J. Matera and Robert F. Metcalf, Jr., *The Leavitt Pumping Engine at Chestnut Hill Station of the Metropolitan District Commission, Boston, Mass.* (Boston, December 14, 1973), 8 pp. illus.; *The A. B. Wood Low-Head, High-Volume Screw Pump, No. 1 Pumping Station* (New Orleans, June 11, 1974), 14 pp. illus.

[34] Marlowe, Donald E., to author, personal communication, Sept. 12, 1974.

Appendix A

Book-length Biographies Published by The American Society of Mechanical Engineers

John Fritz, *Autobiography of John Fritz* (N.Y.: ASME, 1912). 326 pp., illus. Trade edition by John Wiley & Sons. Includes accounts of testimonial dinners given in 1892, when Fritz was 70, and of the banquet "to celebrate the foundation of the John Fritz Gold Medal," held in 1902 (pp. 227-318). Fritz (1822-1913) was a pioneer in the iron and steel industry.

Henry G. Prout, *A Life of George Westinghouse* (N.Y.: ASME, 1921). 375 pp., illus. Trade edition by Charles Scribner's Sons. Prout wrote this book "for a committee of the American Society of Mechanical Engineers," whose chairman was Charles A. Terry. Both Prout and Terry were associates of Westinghouse. Westinghouse (1846-1914) was inventor of the railroad airbrake and founder of the steam turbine and electrical equipment company bearing his name.

Frank B. Copley, *Frederick Winslow Taylor* (2 vols., N.Y.: ASME, 1923). 467, 472 pp., illus. Trade edition by Harper & Brothers. Taylor (1856-1915) was chiefly responsible, both as author and promoter, for the modern scientific management movement.

John A. Brashear; the Autobiography of a Man Who Loved the Stars, William Lucien Scaife, ed. (N.Y.: ASME, 1924), 262 pp., illus. The foreword was written and much of the editing was done by Edna Yost. (See her account of Brashear in *Dictionary of American Biography).* Brashear (1840-1920), prominent maker of astronomical lenses and instruments of precision, was President of ASME in 1915.

Albert W. Smith, *John Edson Sweet* (N.Y.: ASME, 1925). 220 pp., illus. The author was dean emeritus of the Sibley College of Mechanical Engineering at Cornell University. The chairman of the "committee on the Sweet biography" was E. N. Trump. Sweet (1832-1916) taught at Cornell from 1873 until 1879, when he organized the Straight Line Engine Works in Syracuse to build steam engines of his design.

Albert W. Smith, *A Biography of Walter Craig Kerr* (N.Y.: ASME, 1927). 191 pp., illus. Kerr (1858-1910), was a founder of Westinghouse, Church, Kerr & Co., in effect a New York sales organization for the Westinghouse Electric Corporation of Pittsburgh.

This book was published "at the expense of H. H. Westinghouse," brother of George and chairman of Westinghouse Air Brake Company. [ASME, *Report and Index,* I (1927), 126].

Archibald D. Turnbull, *John Stevens; An American Record* (N.Y.: ASME, 1928). 545 pp., illus. Trade edition by The Century Company. Stevens (1749-1838) was a pioneer in steamboats and steam railroads. His son Edwin (1795-1868) left the means to establish Stevens Institute of Technology.

The impetus for this book probably came from A. C. Humphreys, past president of ASME and president of Stevens Institute of Technology, to whom it was dedicated; the author was his son-in-law. *(Mechanical Engineering,* Oct. 1927, p. 1154).

William F. Durand, *Robert Henry Thurston* (N.Y.: ASME, 1929). 301 pp., illus. Thurston (1839-1903) was in 1880 the first president of ASME; he established mechanical engineering curricula at Stevens and at Cornell; his influence on engineering education was both profound and lasting. Durand had taught under Thurston at Cornell. Publication of the volume was assisted by the "Gleason Gift." See Reference Note [19].

L. P. Alford, *Henry Laurence Gantt* (N.Y.: ASME, 1934). 315 pp., illus. Gantt (1861-1910) was a leader in the scientific management movement. The preface of this book quotes large portions of the ASME Biography Committee's specifications as to what a biography "should contain."

Joseph W. Roe, *James Hartness* (N.Y.: ASME, 1937). 147 pp., illus. Hartness (1861-1934) was a prominent machine tool builder, sometime Governor of Vermont and President of ASME in 1914.

William LeRoy Emmet, *The Autobiography of an Engineer* (N.Y.: 233 pp., illus. This is the second edition of Emmet's autobiography, first published privately in 1931. He added two chapters to bring the story through the 1930s. Emmet (1859–1941) was responsible for significant innovations in several classes of electrical apparatus.

Fred J. Miller (N.Y.: ASME 1941). Miller (1857–1939) was an early editor of *American Machinist* who moved subsequently to the typewriter industry. This book, paperbound, consists of the biographical sketch prepared by a committee of five, including Morris Cooke and Fred Colvin, and published in ASME *Transactions*, LXIII (1941), RI72–RI75, supplemented by excerpts from Miller's writings.

Mortimer E. Cooley, *Scientific Blacksmith* (N.Y.: ASME, 1947). 290 pp., illus. Trade edition by University of Michigan Press. Cooley (1855–1944) was a public utilities expert, dean of engineering at the University of Michigan, and President of ASME in 1919.

Edna Yost, *Frank and Lillian Gilbreth: Partners for Life* (N.Y.: ASME, 1949). 372 pp., illus. Trade edition by Rutgers University Press. Frank Bunker Gilbreth (1868–1924) was a pioneer in industrial motion study. Lillian (1878–1973) had carried on her husband's and her own work for 25 years when this book was published.

William F. Durand, *Adventures: In the Navy, in Education, Science, Engineering, and in War: A Life Story* (N.Y.: ASME, 1953). 212 pp., illus. Trade edition by McGraw-Hill. Durand (1859–1958) taught at Stanford until 1924, when he was 65; he was consultant on dams from Hetch-Hetchy to Grand Coulee; he contributed importantly to airplane development. He was President of ASME in 1925.

Dexter S. Kimball, *I Remember* (N.Y.: ASME, 1953). 259 pp., illus. Trade edition by McGraw-Hill. Foreword by Ralph Flanders. Kimball (1865–1952) was dean of engineering at Cornell University.

Appendix B

History References in ASME *Seventy-Seven Year Index 1880–1956* (1957), pp. 58, 248

Many, but by no means all, historical articles are listed in this part of the *Index.* Notice, for example, the historical entry under the subject heading HOISTING MACHINERY. Adams' "Early Hydraulic-Turbine History" and Babcock's "Substitutes for Steam" Reference Notes [23] and [24] will be found under appropriate subject headings.

HISTORY

Development of compound locomotive, by A. Von Borries; Trans. 1893, p. 1172.

History and modern development of art of interchangeable construction in mechanism, by W. F. Durfee; Trans. 1893, p. 1225.

First stationary steam engines in America, by F. R. Hutton; Trans. 1894, p. 982.

Ancient Pompeian boilers, by W. T. Bonner; Trans. 1897, p. 113.

Some landmarks in history of rolling mill, by C. H. Morgan; Trans. 1901, p. 31.

Record of early period of high-speed engineering, by C. T. Porter; Trans. 1901, p. 65.

Brief history of early steam navigation in United States; Jour. 1909, October, p. 23.

John Stevens' part in development of steam navigation; Jour. 1909, November, p. 17.

Nation's call to professional man, by W. H. Taft; Trans. 1917, p. 529; Jour. 1918, p. 5

Carnot centenary commemoration; M.E. 1925, p. 37.

Effect of cotton gin upon politics of United States from 1787 to 1857, by W. C. Saylor; M.E. 1926, p. 1397.

Landmarks in history of prime movers, by H. W. Dickinson; M.E. 1926, p. 1385; correction M.E. 1927, p. 179.

Mechanical invention as form of expression, by Crosby Field; M.E. 1928, p. 447, disc. p. 869.

Engineer as ruler, by H. P. Vowles; M.E. 1930, p. 103.

John Fitch, steamboat inventor, by W. H. Richardson; M.E. 1932, p. 399.

Richard Arkwright and early years of industrial revolution, by H. R. Taube; M.E. 1932, p. 677.

George H. Corliss: one of leading pioneers in development of power generation from Watt to present day, by J. A. Hall and G. W. Richardson; M.E. 1933, p. 403, disc. p. 517.

Lives of engineers, by J. W. Roe; M.E. 1934, p. 8.

Early U. S. steam tables, by A. M. Greene, Jr.; M.E. 1934, p. 715.

Study of technical history, by I. N. Liphshitz; M.E. 1935, p. 143, disc. p. 449.

Technical ancestry of grain-milling devices, by R. H. Anderson; M.E. 1935, p. 611.

Watt, James, 1736–1819, by G. A. Orrok; M.E. 1936, p. 75.

Boulton, Matthew, 1728–1809, by J. W. Roe; M.E. 1936, p. 80.

Engineering achievements of George Westinghouse; M.E. 1937, p. 263.

Langley "whirling table," by J. A. Dent; M.E. 1937, p. 732.

Interchangeable manufacture, by J. W. Roe; M.E. 1937, p. 755.

Benoit Fourneyron (1802–1867), by F. W. Keator; M.E. 1939, p. 295.

HOISTING MACHINERY. *See also* ASH HANDLING, CHAINS, COAL HANDLING, CONSTRUCTION MACHINERY, DREDGES, MATERIALS HANDLING *(yearly progress reports)*, ROPE, WIRE.

History and development of winding and pumping machinery in anthracite regions of Pennsylvania, by Howell Green; Trans. 1883, p. 217.

HISTORY. *See also* BOILER CODE OF A.S.M.E.

Development of equipment for rotary well drilling (1900 to 1940), by J. H. Pound; M.E. 1940, p. 441.

The forge of Vulcan; historical review of the iron industry and the contributions of armorers to the metal-working art, by John Woodman Higgins; M.E. 1941, p. 527.

Pressed-steel industry, by Carter C. Higgins; M.E. 1941, p. 534.

Tale of two city stations; a half century of progress in steam-power generation, by Arthur M. Greene, Jr.; M.E. 1941, p. 109.

Ordnance in my time; forty years of development of American ordnance, by Major General Charles M. Wesson; M.E. 1942, p. 435.

Was Washington an engineer? by George F. Bush; M.E. 1942, p. 130.

Air transports—past and future, by W. W. Davies; M.E. 1943, p. 471.

Centenary of the British Institution of Mechanical Engineers, by R. H. Parsons; M.E. 1947, p. 841.

The engineer and the historian, by Louis Martin Sears; M.E. 1947, p. 581.

History and exposition of the laws of thermodynamics, by Joseph H. Keenan and Ascher H. Shapiro; M.E. 1947, p. 915.

Preserving the Cornish engine, by James T. Kemp; M.E. 1947, p. 365; disc. M.E. 1947, p. 867.

Sir Charles Parsons and the "Turbinia," by H. Philip Spratt; M.E. 1947, p. 585.

Who was Holley? M.E. 1947, p. 564.

The original Froude hull models, by H. Philip Spratt; M.E. 1948, p. 136.

Nature of fatigue of metals, by R. E. Peterson; M.E. 1950, p. 371.

Plywood is engineered wood, by Thomas D. Perry; M.E. 1952, p. 787; disc. M.E. 1953, p. 247.

Woodworking machinery—History of development from 1852–1952, by Judson H. Mansfield; M.E. 1952, p. 983; disc. M.E. 1953, p. 504.

Adam couldn't fly, by George H. Prudden; M.E. 1953, p. 896.

Gas turbines in our century, by Claude Seippel; Trans. 1953, p. 121.

A.S.M.E. role in powered flight, by Harris F. Reeve; M.E. 1953, p. 987; disc. M.E. 1954, p. 373.

Early history of the frequency-response field, by A. C. Hall; Trans. 1954, p. 1153.

Physics—Mother of instrumentation, by Gaylord P. Harnwell; M.E. 1954, p. 987.

Mechanical-engineering progress in the petroleum industry—Review of developments, by Emory N. Kemler; M.E. 1955, p. 1042; disc. M.E. 1956, p. 467.

Electric-power generation—Past, present, and future, by J. B. McClure and A. G. Mellor; M.E. 1956, p. 521.

Outdoor-type steam-electric stations early types and development, by Louis Elliott; Trans. 1956, p. 671.

Review of the metal-cutting analyses of the past hundred years, by I. Finnie; M.E. 1956, p. 715.

The cable machinery in the Power House of 1887 at Washington and Mason Streets. Electric motors have replaced the original steam power, although the stack remains as a reminder of the past.

Ferries and Cliff House Cable Railway—1887

Washington and Mason Streets
San Francisco, California

Designated
November 30, 1973

Cable cars and San Francisco have long been synonymous. The cars and the system are symbol and hallmark of the City on the Great Bay; many of the cars in daily operation are older than the earthquake and fire of 1906, some dating from 1886. In 1890 some 600 cars operated over 110 miles (176 km) of single-cable track, and there were a couple of horse lines and a steam road as well. Of all that, only 10 miles (16 km) of single-cable line have survived, but these still represent much of the soul of the town.

It was in 1869 that Andrew Hallidie, a manufacturer of wire rope, came upon a heavily laden horsecar trying to make it up a hill. A horse stumbled and fell, the brake failed, and four frantic horses were dragged down hill by the runaway car. Hallidie schemed up a transportation system in which a car could be attached to an endless cable running continuously in a cavity or conduit between rails in the street.

In 1872 a corporation to build a line up the Clay Street hill was formed. The design of the cable grip, the car's vital element that reached down to the under-street cable through a slot in the street itself, was a major problem. With the cable running at 9 miles per hour (14.5 km/h) on Clay Street's 20 percent (11.5 degree) grade, the successful trials of 1873 had the first cable line in operation in the United States (a primitive forerunner had come into existence in London 37 years earlier, and New Orleans had an experimental line in 1870).

The 1870 s saw the transition from animal to mechanical power – it was the era of the cable railway. The superiority of cable traction over the horsecar saw the spread of the technology to every major American city except Detroit and Boston, Chicago being the first installation outside of San Francisco. Overseas, comprehensive cable systems were found in only Edinburgh and Melbourne. However, except for special situations such as very steep grades, cable traction was technologically obsolete by 1888, electric street cars offering much simpler power distribution: the trolley wire was incomparably cheaper than cable traction, whether the latter was the deep-conduit California system or the later and less expensive shallow-conduit system (Tom Johnson) that was aborted with the introduction of electricity. During the last quarter of the nineteenth century, street railways, from the horsecar on, were responsible for the extension of city limits and the formation of suburbs.

Hallidie's convincing demonstration on Clay Street quickly led to further cable car franchises. By 1890 San Francisco had eight competing cable line companies, and three noncable companies using horse- and steam-power. In the 1890s consolidations reduced the number of companies to five. In much more recent time, the number of lines dropped to three, all owned and operated by the city's municipal transit system. These lines constitute the "Ferries and Cliff House Cable Railway" designed in 1887 by Howard C. Holmes (1854–1921), a civil engineer who was also responsible for cable-car systems in Portland and Seattle. The heart of the present operation and the home of the cable cars is "The Barn," or officially, the Washington-Mason Car and Power House. The original three-story building of 1886, the car house, and most of the cars and equipment, were almost completely destroyed in the catastrophe of 1906.

Today's "Barn" is the reconstruction of 1907, whose tall stack for the old steam power-plant serves as a marker of the structure now undergoing cosmetic treatment to bring it into harmony with the venerable brick and gaslight atmosphere of early San Francisco. The Corliss steam engines – two 500 hp (373 kW) vertical and two 450 hp (335 kW) horizontal – have given way to electric power, motors of 700 and 750 hp (520 and 560 kW) alternating with each other. That the system is complicated is evident from the fact that four cables of lengths varying from nearly 2 to 4 miles (2.8 to 6.6 km) are necessary for the present three cable car lines; all cables focus on the "Barn." The entire system has 10.7 miles (17.2 km) of cable railway of 3.5 ft (1060 mm) gage.

The 1¼ in. (32 mm) plow steel cables running 18 in. (460 mm) below the street level and supported on pulleys at about 32 ft (10 m) intervals have an ultimate strength of

The look of things on the first line in 1883.

155,000 lb (690 kN), and last between three and ten months depending upon location, season, number of cars and passenger volume. The splice making the cable endless is some 66 ft (20 m) long – six tucks 11 ft apart; the long splice must finish to not more than the diameter of the cable so that there is no lump for the "grip" to foul on.

The grip is a pincer-like affair that grabs onto the moving cable when the "gripman" (who has to be a rather husky fellow) hauls on his lever. The pincer-jaws are lined with replaceable foot-long soft-steel "dies" introduced in 1922 that can put a 1500 lb (680 kg) squeeze or grip on the running cable to bring the 6 ton (5.45 t) car into motion. Even though the cable is lubricated with a pure Louisiana pine tar, the dies have a four-or-five-day life; the short cable-life can also be understood at this point. When clamped onto the cable, the cars move at the 9 mph (14.5 km/h) cable speed whether going downhill or up.

The speed of even a descending car is limited to the cable's steady 9 mph, for the cars are separated from the cable only at the moment of a deliberate stop. For such stops there are conventional wheel brakes. These are backed by track brakes, wooden blocks about 2 ft long and 3 in. wide (fitted between the wheels of each truck) that can be pushed down onto the track, tending to lift or support the car. As a last resort for crisis conditions, there is the emergency or slot brake, a tapered piece of metal 18 in. wide that can be jammed into the cable slot to wedge the car to a halt.

San Francisco's local convenience and great tourist attraction that can boast a century of operation without a fatality on the hills of the Bay City is practically unchanged from the 1880s with the exception of the cable power, electricity having taken over for steam. Most of the cars are of vintage dates; there was a slight change in grip design in 1956; storage batteries and headlights were added at some time; electric windshield wipers in 1964 represent the ultimate! For all practical purposes, today's cable car passengers are riding in and with the past.

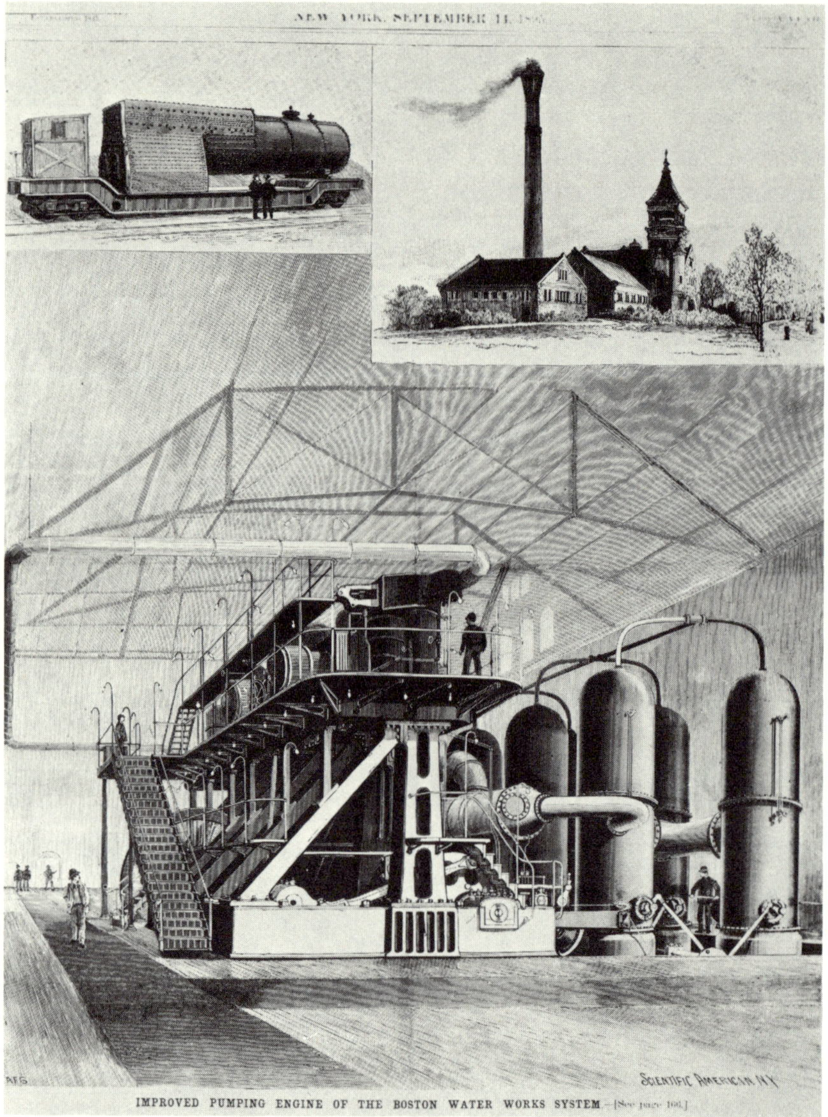

The Leavitt engine montage from Scientific American

Leavitt Pumping Engine—1894

Chestnut Hill Pumping Station
Brookline, Massachusetts

Boston appears to have been the first city to have its water supply remarked upon, since in 1652 the Water Works Corporation constructed a reservoir 12 feet square fed by wooden pipes stretching out to nearby springs. But between then and 1894, nearly two and one-half centuries later, when the Leavitt Pumping Engine known as No. 3 was installed, too much has passed for the telling.

Engine No. 3 is of unusual design, and for its day, was a high-capacity unit that attracted much attention because of outstanding performance. It served for 34 years before going onto standby status in 1928 and for some years thereafter. The nameplate identifies it as a "Riedler Pumping Engine, Designed by E. D. Leavitt"; the explanation will come in course.

The engine is a somewhat unconventional triple expansion machine of 6 ft (1828 mm) stroke with cylinder bores of 13.7 in. (347 mm), 24.375 in. (619 mm) and 39 in. (990 mm). The cylinders are vertical and inverted, inversion meaning that the piston rods emerge from the lower end of the cylinder instead of following the tradition established by Newcomen in 1712 and followed by Watt and many other designers in which the piston rod came out of the top of the cylinder to reach up. Both cylinders and valve gear are carried on an entablature supported by six vertical columns and six diagonal members rising from the bed plate.

Each connecting rod, instead of terminating at a crank that can rotate, drives to a rocker from which one link does go to a crank on the shaft of a 15 ft (4560 mm) diameter flywheel, a second link becoming the connecting rod of a reciprocating pump.

Designated
December 14, 1973

The crankshaft has three cranks at 120 deg, the low pressure crank leading, followed by the intermediate and high pressure cranks.

Three double-acting inclined pump plungers have a 4 ft (1200 mm) stroke and 17.5 in. (444 mm) bore. The pump valve mechanism, somewhat similar to the valve-rod system of a Corliss engine, one of the characteristic features of the unit, is based upon the invention of Austrian-born Professor Riedler of the Royal Institute of Technology, Berlin-Charlottenburg. The Riedler pump valve-gear, introduced to this country by Leavitt in an earlier pumping unit (Louisville, 1893), gave quick and positive valve action for the large pump volumes at the then high speed of 50-60 rpm. A normal speed of 50 rpm had a 20 million gal (75700 m^3) per 24 hour capability against a head of 128 ft (35.6 m), the engine developing about 575 hp (428 kW) on steam at 175 psi (1205 kPa) with 27 in. (690 mm) vacuum in the condenser.

The unit's designer, Erasmus D. Leavitt, Jr. (1836-1916) was a man of great ability and distinction. He became a machinist apprentice at age 16, and when 22 years old, was put in charge of the construction of the engine for the USS HARTFORD, the Union flagship in the brilliant battle of Mobile Bay. He enlisted in the Navy in 1861 and left it in 1867 after having been instructor in Steam Engineering at Annapolis for some years. In private practice as mechanical engineer, he specialized in pumping and mining machinery, acquiring 16 patents in 34 years. An efficient beam pumping engine of his design installed and tested at Lynn in 1873 brought prominence and acquaintance, including the contact with Riedler from whom he gained the rights for the pump and valve gear in the U.S. On the recommendation of James B. Francis, the hydraulic engineer of Lowell, Massachusetts, Leavitt was made consulting engineer to the Calumet and Hecla Mining Company, an association lasting 30 years. His reputation and fame grew, and he was recognized as having done more than his contemporaries to establish sound principles and propriety of design. He objected to excessive decoration and embellishment of machinery, saying that the beauty of a machine was determined by its practicality and fitness for the job.

Among his many honors was the honorary doctorate from Stevens Institute of Technology in 1884, the first awarded by that institution. He was a founding member of ASME, and became President in 1883, its third year.

This 14-foot pump was built in 1929.

A. B. Wood Low-Head High-Volume Screw Pump—1913

Melpomene and Broad Streets
New Orleans, Louisiana

From its beginning, New Orleans suffered from drainage-related epidemics and visitations of yellow fever and cholera, and always malaria; the city was known as one of the most unhealthy spots in North America. New Orleans soil is an alluvial deposit, the greatest elevation being at the river bank, the ground behind the levees having built itself to an elevation of 10-15 ft (3 to 4.5 m) above the mean level of the Gulf of Mexico, whereas the interior parts of the city are 4 – 12 ft (12 to 3.6 m) below the Gulf level. Over the years levees and dikes have encircled the city, and, while keeping the river and lake out of town, these water defenses also retain local waters, such as sewage, industrial wastes and rain (60 in. or 1.5 m annually). Sewage and runoff have to be disposed of mechanically, i.e., lifted or pumped over the "city walls" of levee and dike.

Early plans to supply the city with pure water, adequate drainage and proper sewage disposal, although proposed from time to time, were impotent because of lack of financing. Until the 1900s practically all of the drainage – rainfall and whatever gutter-thin refuse would flow into open canals representing a token drainage system – was handled by four inadequate steam-driven paddle-or scoop-wheels forcing the polluted waters into an adjoining swamp. Drains were clogged, the water table was very high, and the open canals and city swamps bred mosquitoes and disease. Cesspool refuse was simply dumped into the river from which the drinking water was taken. In defense, householders collected rain water in great cypress cisterns for domestic purposes, but the water became stagnant and also bred mosquitoes.

Designated
June 11, 1974

In 1896 the Lousiana legislature created the New Orleans Drainage Commission that drafted plans and began construction of functional drainage and sewage systems. With an

The first and still active "Wood 12-foot Screw Type" pumps and electric motors installed in Drainage Station No. 1 in 1915. Some of the blading is visible.

apathetic public, funding was limited until the recurrence of yellow fever in 1898 and 1899 aroused support and saw the creation of the Sewerage and Water Board (S & W B). By 1900 progress was in hand. In particular, vertical-shaft screw pumps of 8 ft (3400 mm) diameter driven by synchronous motors were now doing the pumping. (A screw pump is a rotary machine having an impeller with a single row of twisted blades that are short sections of a thin helix or screw thread; if there are eight blades, one may think of an aborted eight-start screw. A screw pump is an axial-flow machine.) The design of these vertical pumps placed them at the low points of the system, making servicing difficult and nasty when they were drowned due to malfunction.

The need for increased drainage pumping facilities was pressing. A young mechanical engineer of the S & W B, A. Baldwin Wood, designed the first of a series of horizontal-shaft screw pumps that would gain an international reputation as "Wood Screw Pumps," machines for low-head but high-volume water problems, and placed at the readily accessible highest point of the system.

These first pumps were located at the top of the siphon between the low water or suction basin and the several-feet-higher discharge. By 1915 four of the thirteen pumps that had been ordered had been installed, and a performance test run on one. The report here speaks for itself: "The pump is larger than any centrifugal pump ever built and among the largest screw pumps, 12 feet (3650 mm) in diameter, of horizontal type, designed to give 247,500 gallons per minute (937 m^3/min) against a 7 ft (2.1 m) lift at 75 rpm and to work at this constant speed driven by a 600 hp (450 kW) motor for any lift from 0 - 10 ft (0 to 3 m) . . . . Results showed that the pumps easily answer all requirements and that they are the largest and most efficient low-lift pumps in the world." Today's drainage system consists of Wood Screw Pumps of various sizes up to 14 ft (4240 mm) diameter. Following their successful performance in New Orleans, similar pumps were installed in the Netherlands, Egypt, China and India.

A. Baldwin Wood (1879-1956) was a native son, one ancestor having been Governor and Administrator of the Province of Louisiana in the late eighteenth century. Wood, a Tulane University graduate who joined the S & W B at the age of twenty, rose to the position of General Superintendent. His activity is attested to in part by 38 patents. He was a consultant to public and private agencies, and among other things redesigned Chicago's drainage system.

The first Portsmouth submarine, the L-8, under construction in the Franklin Shiphouse in 1915.

Portsmouth-Kittery Naval Shipbuilding Activity—1774

Portsmouth Naval Shipyard
Portsmouth, New Hampshire

Designated
March 22, 1975

The mouth of the Piscataqua River between Maine and New Hampshire has a good and deep harbor whose singular advantages were recognized in earliest colonial time and soon exploited. The stream itself was "of great depth with beautiful islands and heavy forests along its banks." Forests were vital to maritime nations, since oak was used for hulls, pine for masts, fir for yards, and poplar for ornamental carvings. In seventeenth-century England all oak trees were reserved for the Royal Navy; augments to the native supply were necessary and came most easily from the American colonies and the West Indies. New World timber for masts and spars was equally essential. So it came that the great up-river forests occasioned Portsmouth long before the several islands were organized to build ships for the Royal Navy, as the three frigates having 54, 32 and 60 guns constructed at Portsmouth in the years 1690, 1696 and 1749.

In colonial days, the present Fort Constitution (in New Castle, New Hampshire) commanding the harbor entrance since 1690, was known as Fort William and Mary, having been named for the rulers of England who succeeded James II in 1689. In December 1774 the six-man British garrison of the fort was overpowered by the local colonists who were reacting to Paul Revere's ride to Portsmouth of the day before. The Continental Congress, it might be mentioned, was not yet contemplating forcible resistance to England, having indeed proposed measures to "restore harmony." This seizure of the fort was the first organized military action against the Mother Country. The captured supplies — powder and cannon — were dispersed here and there. Some of the powder was later used in the traditional first engagement of the War of Independence, the skirmish at Lexington in April of 1775.

With the Piscataqua free of British forces, the Portsmouth-Kittery Naval Shipbuilding Activity that is part of the present one got under way, although the official "Navy Yard" in the vicinity was not established until 1800. In December 1775 the Continental Congress, wary of the progress of the negotiations with England, authorized the construction of thirteen frigates, to be ready in three months. Among these ships was the 32-gun RALEIGH to be built at Portsmouth. This contract, the first of many, was completed successfully, the ship being launched sixty days after the keel was laid, and six weeks before the Declaration of Independence.

Portsmouth's second ship for the Continental Navy, the 18-gun sloop RANGER, was built on the same blocks that had carried the RALEIGH's keel. Launched in May 1777, this ship, under the command of Captain John Paul Jones, may have been the first vessel to carry the new flag of thirteen stars and thirteen stripes to Europe, sailing from Portsmouth in December, 1777. The first American ship-of-the-line, the 74-gun AMERICA, belongs also to the roster of Portsmouth-built naval vessels.

With time, Portsmouth became a fully integrated shipbuilding operation, a truly self-sufficient industrial complex. It had its own foundry, forge- and blacksmith-shops, shops for tin- and coppersmiths, woodworkers, and plumbers, as well as rope walk, mast shop and rigging and sail lofts. It grew some of its own food, even to the keeping of a small herd of cattle.

Of the present thirty-odd buildings constructed before the year 1900, about half predate the Civil War. The oldest building, from 1837, is the "Mast and Boat Shop and Riggers and Sail Loft"; obviously its present activity is somewhat less than that of the description. Of rugged construction, the walls are granite. The timbers show the marks of axe and adze, and are fastened with mortise and tenon joints and treenails. Originally an open building, there was a canal down the center by which logs could be floated in and converted into masts and spars after being winched up to the first-floor shop. Some of the mast gear is still at hand. Two floors were added in later years, and the building is in use as the Riggers Shop.

Portsmouth built many ships for the Navy in moving from sail to steam during the nineteenth century. The first steam vessel was a side-wheel frigate launched in 1848.

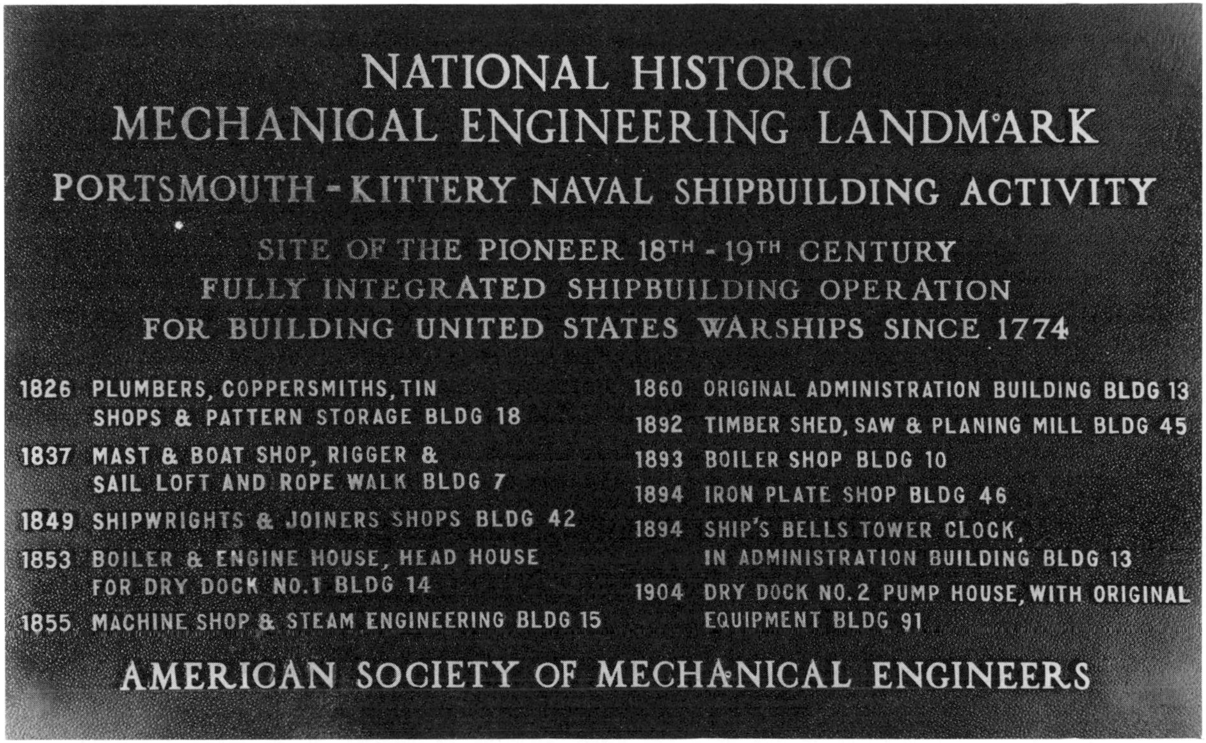

The plaque indicates the principal "mechanical" operations

Today "old" Portsmouth is extremely modern, having been the first government Navy Yard to build a submarine, the L-8, whose keel was laid in November 1914. The Yard – since 1963 the official name is Portsmouth Naval Shipyard, even though it happens to be on an island belonging to Kittery – is now a center for the building and repair of submarines. It is the first government yard to have constructed a nuclear submarine, as for example, the SEADRAGON that in August 1962 rendezvoused under the North Pole with the SKATE.

One of the Boyden turbines of 1871–1873, with section of overhead main pulley line-shaft. The bevel-gear drive originating on the vertical turbine shaft is hidden by the circular shrouds. The water entered the turbine from the back wall and made a 90-degree turn before hitting the blading of the runner at a level below that of the floor.

102 Inch Boyden Hydraulic Turbines—1871/1873

Harmony Mill No. 3 on North Mohawk Street
Cohoes, New York

Designated
May 28, 1975

During the second quarter of the 1800 s waterpower moved from the bulky, slow-turning waterwheel to the much more compact and speedier water turbine of superior efficiency. Barker's reaction wheel and its improved version, the three-orifice Scotch wheel, were followed by the many-jetted wheel. To this Fourneyron fitted fixed guide blades just inside the wheel ring to create the first modern turbine in 1826, a radial, outward-flow motor that would be varied and improved upon by many. In about 1845 Uriah A. Boyden (1804–1879) improved the performance of the Fourneyron turbine by providing a conical approach passage, inclined vanes and a submerged diffuser. In time, the outward flow Boyden turbine would be superseded by the Francis type, an inward flow machine.

The Harmony Mill No. 3, also known as the Mastodon Mill, an architecturally elaborate component of a large textile manufacturing company, was the site of several Boyden wheels of which two remain in situ as historic entities. They are reminders of the time before the electrification of 1915 when Cohoes, like many Eastern manufacturing towns, depended upon water for industrial power. At Cohoes, located on the former "Great Falls" of the Mohawk that flowed into the Hudson, a dam diverted water to mills and factories by a power canal system that gave a 70 ft (21.3 m) head.

In the early 1800 s a textile mill of any consequence was powered by an in-house waterwheel, a breast wheel of substantial diameter and considerable length. For example, even the modest Wilkinson mill in Pawtucket had a wheel of 12 ft (3600 mm) diameter and 13 ft (3900 mm) length. Such a slow-turning waterwheel would drive cast iron shafting at higher speeds by means of spur and bevel gears. Vertical shafts carried the power from story to story, feeding overhead lineshafts to which the machinery was belted. Wheel efficiencies were of the order of 60 percent, and all that cast iron meant a massive setup, even the pulleys being large because of the relatively low speed imposed by the shafting of square or diamond section except where journaled.

The modern note was struck by an installation such as the one at Cohoes. The power – now relatively high-speed, vertical-shaft turbines of over 80 percent efficiency, as our Boyden wheels – was still in the cellar, and drove an overhead main pulley line. From this, power was distributed as needed by means of flat belts of sometimes impressive widths and lengths. The horizontal line shafting was wrought iron or even steel, of circular section and smaller than the cast iron it supplanted. The now-possible higher shaft speeds permitted smaller and lighter pulleys.

The two Boyden hydraulic turbines recognized as Landmarks were possibly among the largest and the nearly most powerful water motors built in the United States for driving a manufactory. Installed sometime between 1871 and 1873, they are among the oldest surviving mill turbines. Although the turbines are still in place at the south end of the mill building, surviving data relating to them are few. The runners have a diameter of 102 in. (2600 mm); the horsepower of each is said to have been 800 (600 kW) when operating under an estimated head of 20 ft (6.1 m) at an unknown speed. The two vertical-shaft turbines were connected to a common overhead horizontal shaft by bevel gearing. Belts on large pulleys of the common shaft reached up to line shafts on the floors of the five-story building. This shafting installation no longer exists, but it must have been magnificent, especially the belt (not rope) drives, the longest of which may have been of the order of 85 ft (25.9 m) center-to-center.

Some scaled drawings of the north end of the mill building have been preserved and are worth examination for the kindred information they provide. At the time of construction, 1866-1868, three Boyden turbines, of unknown power but smaller than the two Landmark motors, were installed to drive the machinery in the north end. These turbines were scrapped at an unrecorded date.

From the drawings it is seen that the three turbines were bevel-gear-connected to a common overhead horizontal shaft about 85 ft (25 m) long, giving it and its six pulleys

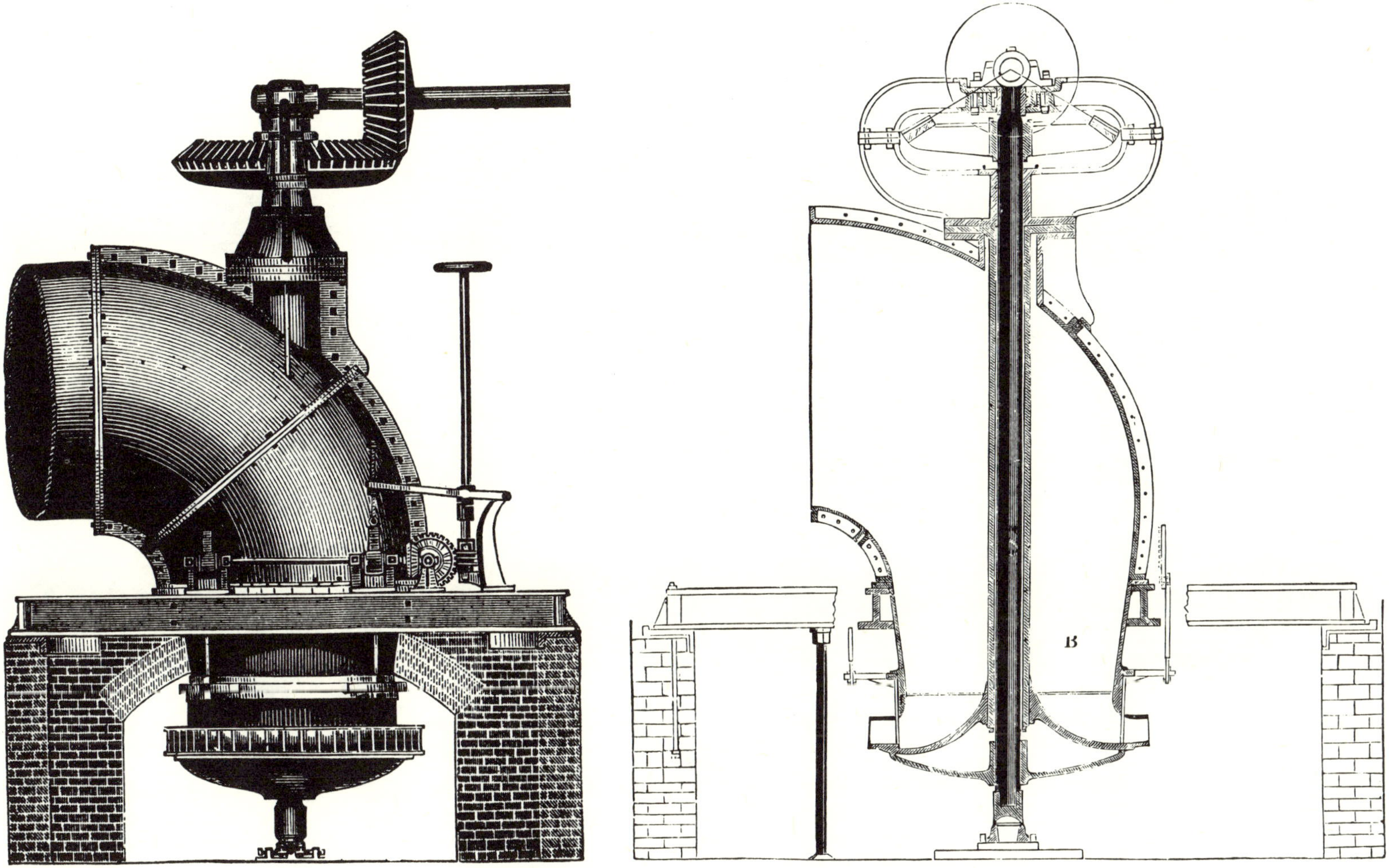

Not the surviving machine but a similar one shown in the manufacturer's catalog of 1876.

Harmony Mill, No. 3, Cohoes, N.Y.

Scale of Feet.

50 Feet 40 30 20 10 5 0

Section of Mill showing Turbines and Belting. **Part Plan of Mill and Turbines.**

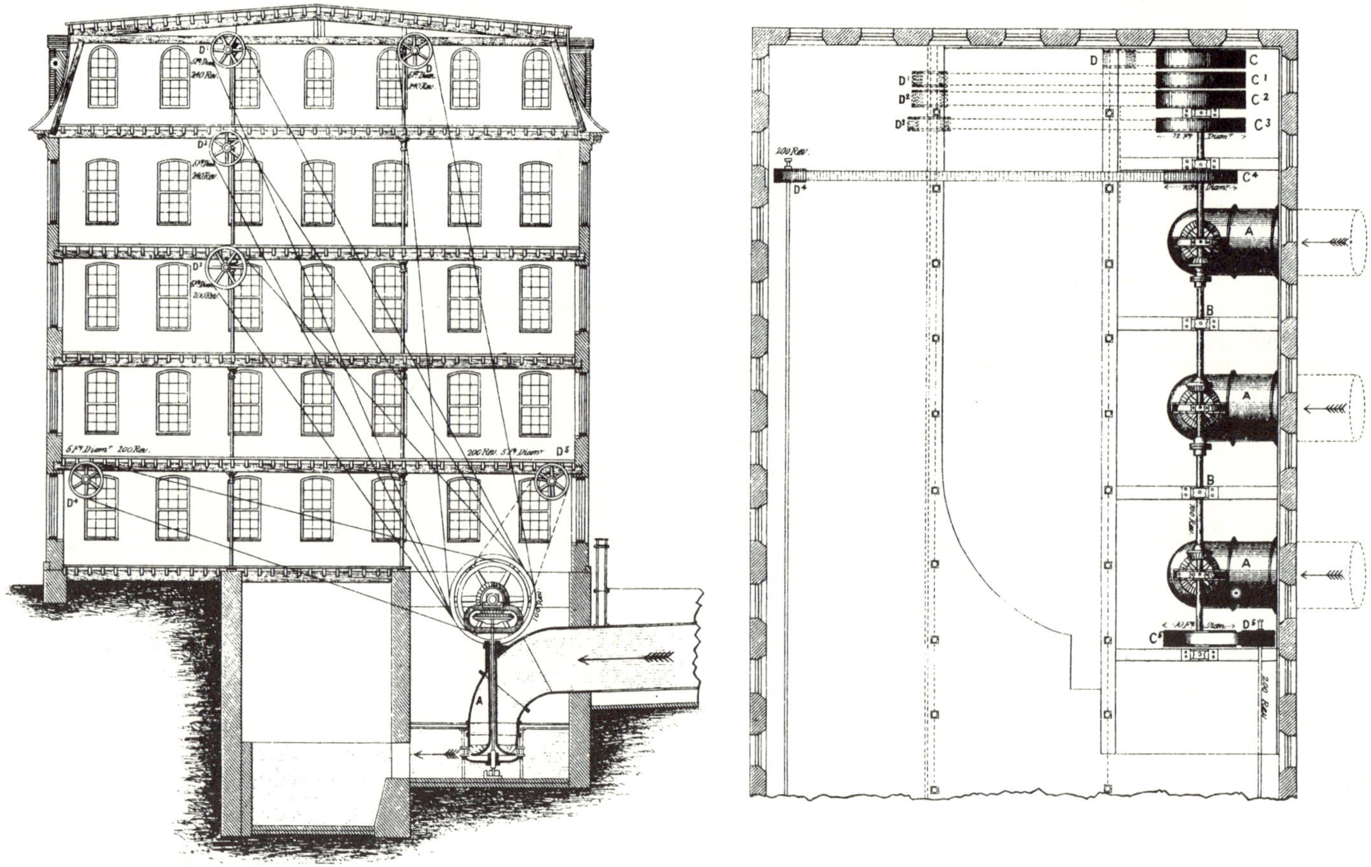

The machine/power-transmission arrangement at the north end of the mill, mentioned in the text.

a speed of 100 rpm. From the bevel gear diameters the turbine speed may be estimated to have been about 60 rpm. Four of the pulleys were of 12 ft (3650 mm) diameter, while two were but 10 ft (3050 mm). Depending upon the location, the lineshafts on the several floors were driven at either 200 or 240 rpm. Center-to-center distances of the shafts ranged from 18 to 85 ft (5.5 to 26 m) and if the drawing is accurate, the belt widths scale out at 12 and 18 inches (305 and 457 mm). The number of cattle hides in the belts must have been astounding. The installation that was at the north end of the mill is probably representative of the situation at the south end with two its remaining 800 hp (600 kW) turbines.

Uriah Atherton Boyden (1804-1879) was born and died in Massachusetts. Dealing with various aspects of civil and mechanical engineering, he became associated with the Amoskeag Manufacturing Company for which he designed a hydraulic power system at Manchester, New Hampshire. In 1844 he built the improved version of the Fourneyron turbine to create the motor that would carry his name. This 75 hp (56 kW) machine, installed at the Appleton Company mill at Lowell showed the pleasing efficiency of 78 percent on test. Several other machines were built and tested by J. B. Francis who was stimulated to modify the Boyden turbine to such an extent that the redesign would carry his name. Boyden is reputed to be the inventor of the hook gage used for measuring the head at a weir from which the flow through the notch may be calculated.

James Bicheno Francis (1815-1892), born in England, emigrated to the United States in 1833 and found employment as assistant engineer on the construction of the Stonington and Providence Railway. Joining next the Locks and Canal Company in Lowell, he became Chief Engineer in 1837 and remained with the firm (in one capacity or another) for the remainder of his life. The radial, inward-flow turbine, generally known as the Francis turbine, was based upon the tests of Boyden wheels, published as the *Lowell Hydraulic Experiments*, 1855. These tests have been a model for hydraulic testing and represent an outstanding American contribution to the art. Francis was a founding member of the American Society of Civil Engineers, and served as President in 1880.

The turbine-generator on display at Building 60 of the General Electric Company about 1910. Left to right, E. W. Rice, Jr., W. L. R. Emmett, Thomas A. Edison, George F. Morrison, Charles P. Steinmetz, and H. F. T. Erben. The machine is now outside the Turbine-Generator Development Laboratory, Building 263. Photo courtesy of General Electric Company.

5000 kW Vertical Curtis Steam-Turbine Generator—1903

General Electric Company Plant
Schenectady, New York

Designated
May 28, 1975

In the year of its construction, 1903, this 5000 kW unit was unique. As the most powerful steam turbine of the world – the next most powerful seems to have been a 1500 kW unit – it represented a significant advance in turbine capacity. Occupying a much smaller space than a reciprocating engine generator of like capacity, it stood but 25 ft (7.6 m) high to the engine's 60 (18.3 m), and required but a fraction of the engine's floor area. Moreover, the compact high-speed turbine cost much less than a commandingly massive vertical-cylinder steam engine driving a 40 ft diameter (12.2 m) generator. Then, too, the large reciprocating engines, slow-speed though they were, produced vibrations of intensities often great enough to disturb neighbors, whereas the rotating turbine elements are easily brought to a state of nearly perfect balance. Finally, the inlet and exhaust conditions of a turbine's power cycle make for efficiencies impossible to attain with a reciprocating engine. Steam at very high pressure and temperature is acceptable to the turbine requiring no interior lubrication, whereas the oil needed for pistons and valves would be carbonized in a cylinder machine given the same inlet steam. Moreover, the expanding steam flowing continuously through a turbine is never confined to a finite cylinder volume, allowing a lower exhaust pressure with modest overall machine dimensions, while an engine would require impossibly large cylinders to achieve the same expansion, expansion being the key to economy.

At the turn of the century, Chicago had an imaginative and farsighted utility executive, Samuel Insull, president of the Commonwealth Electric Company (now Commonwealth Edison Company). He was responsible for what was probably the world's first steam turbine electric generating station – the Fisk Street Station, 1903. Into it he

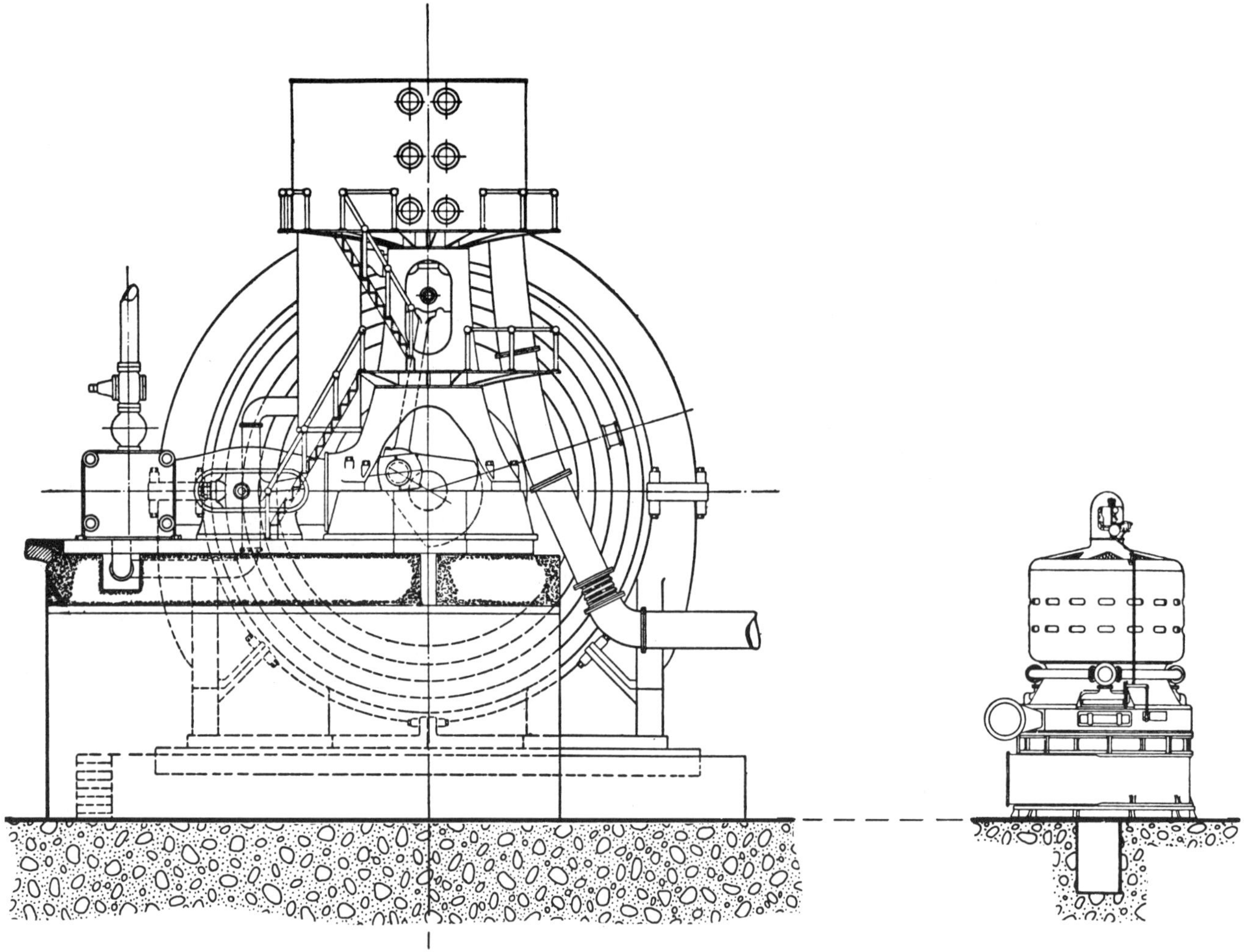

Comparison of 5000 kW machines—a vertical Corliss reciprocating steam engine-generator and vertical Curtis steam turbine-generator.

brought the most powerful turbine generator of the day, a 5000 kW (6700 hp) vertical Curtis Steam Turbine Generator built by General Electric, producing 9000 v, 25-Hz, 3-phase power. The steam was 175 psi (1205 kPa) with 150° F (65° C) superheat exhausting to a 28 in. (710 mm) Hg vacuum at the outlet. This move, and the success of this machine, were of special significance in the history of electrical power generation for it spelled the end of magnificent but also cumbersome reciprocating engine generators, and spurred the growth of ever-larger turbine units of increasing efficiencies. Soon flanked by two similar machines, all three were soon replaced (1909) by improved Curtis machines of 12 000 kW, the first 5000 kW unit returning to Schenectady to assume its honored place in the General Electric Works to also become landmark.

The Curtis vertical machines include the ideas of two men, C. G. Curtis (1860-1953), and W. LeR. Emmet (1859-1941). The thermodynamics – technically, the machine is a velocity compounded impulse type – stem from Curtis' patent of 1896, the manufacturing rights for land use being acquired by General Electric in 1902, Curtis reserving to himself the area of marine applications. General Electric's early Curtis generators were vertical-shaft machines, the generator being above the turbine. This configuration, due to Emmet, was commonly used with hydroelectric generators, but abandoned a few years later as the speeds increased from the 500 rpm of the Landmark to 1800 and then 3600 rpm. While the vertical shaft was feasible for small and slow-speed machines, larger and higher-speed units would require stiffer structures needing to invoke lateral support from the power station building itself. Additional stages for more power allowed higher steam inlet pressures that improved the efficiency but gave shaft lengths that were impractical for the vertical configuration. A horizontal shaft machine draws upon the building foundation for heightened local stiffness.

Saugus Ironworks as they might have appeared about 1650. The ironmaster's house (upper left) is the sole survivor of the original structures and is one of the oldest homes left in New England. Note the several waterwheels; the blast furnace (center) hides another.

Saugus Iron Works—1647

Saugus, Massachusetts
(Ten miles north of Boston)

Designated
June 15, 1975

England's Parliament, when dealing with colonial matters, would ordinarily have forbidden the development of a local manufacturing industry such as the iron works that were permitted in the earliest American colonies. But since the making of iron required great quantities of timber in the form of charcoal that England could no longer supply without sacrificing naval supremacy and trade expansion – ships and their ironwork claimed wood at a rate greater than that at which it could be grown – the colonies were allowed self-sufficiency in at least this department in the seventeenth century. However, there would be change. By the Iron Act of 1750 Parliament forbade the building of new forges and rolling and slitting mills. This prohibition, plus many others, would underlie the Revolution.

The first American iron ore to be discovered, on a North Carolina island in 1585, seems to have been too inconveniently located to be exploited. Ore from deposits along Virginia's James River, found in 1607, was sent to England in the following year, an arrangement quickly discontinued. By 1619 it was planned to build three ironworks in Virginia that would use the readily mined surface ores. The first works was sited at Falling Creek below Richmond. Just as it was to go into operation in 1622, the settlement fell to an Indian massacre, the only survivors of twenty-nine being a boy and a girl. No attempt was made to go on.

The scene moved north, to the Massachusetts Bay Colony, where in 1644 the village of Hammersmith on the Saugus River near Lynn became the first successful iron works and the first manufacturing establishment in the American colonies. Two kinds of iron were produced – pig iron that could be cast to shape, and bar or wrought iron that could be shaped only by forging.

Pig iron was suited to hollow ware such as pots, pans, and skillets, as well as stove parts and other cast articles in which brittleness could be tolerated. Pig iron was also the source of wrought iron, for on being treated in the finery excess carbon was burned out, leaving a malleable material that was forged into bar stock after reheating in the chafery, from which tools, nails, bolts, horseshoes, wagon tires and implements of many kinds were made.

Saugus, a large-scale enterprise in its day, was a completely integrated ironworks, one whose equipment and operation competed with the best contemporary establishments of Europe. For all of this, however, there were financial difficulties, suspicions of mismanagement or worse, raw material shortages, inability to meet the competition of later imported iron, trouble with the local authorities, and the difficulty of obtaining and keeping skilled workers in a new settlement where land was cheap and labor dear. The downhill operation ground to a stop in the 1670 s; by 1678 the works had been abandoned and the local residents were asking that the dam, flooding a thousand acres, be cleared away to improve the fishing. Nature took over, and the ironworks became a collection of grassy mounds but for the iron-master's house which had been built in about 1646.

That there is an authentic recreation of the great undertaking of the 1600 s is due to the American Iron and Steel Institute that had the wit to recognize the historical and educational significance such an exhibit would have. Some six years of work and a million and a half dollars were needed. There were no plans nor designs, and the local records were mostly gone. Fortunately, foundation outlines are permanent unless the stones are carted away by the locals. The reconstruction was opened to the public in September of 1954. In more recent time, the new Saugus has been designated a National Historic Site and is administered by the National Park Service.

It is clear from the various records that Hammersmith – Saugus – was conceived as an integrated whole: the unity is indicated by the layout for material flow. Saugus was "modern," following the best practice of the time. The principal units are three – furnace, forge and rolling/slitting mill.

An earth dam, 18 ft (5.45 m) high, 100 ft (30.4 m) long, and 75 ft (22.7 m) thick at the base, completely lined with stone on the river side, backed the Saugus River into a thousand-acre pond from which a 1600 ft (485 m) canal led to a standby reservoir close to the plant. Several flumes fingered out from this second pond to feed the various water wheels working bellows, trip-hammer and rolling mill.

The furnace, a truncated pyramid of granite and local stone, is 26 by 26 ft (7.9 by 7.9 m) at the base and 21 ft (6.35 m) high with inward sloping walls. A fair portion of the original furnace waterwheel was found buried 22.5 ft (6.8 m) under Central Street, a main traffic route now, but an area that had been the city dump at one time. The re-creation is a six-spoked undershot wheel of 16 ft (4850 mm) diameter and 2 ft (609 mm) face mounted on a shaft made from a Maryland oak tree 125 years old. It actuates, by means of cams on its shaft, two 18 ft long (5.4 m) bellows. There is a suggestion that the furnace might have furnished 500 lb (227 kg) of iron every 24 hours, but the actual figure is uncertain. An estimate – there are no company accounts – has a charge of 3 tons (2.7 t) of bog ore, 2 tons (1.8 t) of flux, and 265 bushels (9.3 m^3) of charcoal accounting for 1 ton (0.9 t) of iron.

The fluid cast iron became either a finished-product casting as pot or pan, or a rough slab-like casting called sow if large, and pig if small, the base stock for making the malleable wrought iron for tools and implements. This transformation would be carried out in a pair of hearths worked in tandem in the Forge House, following the so-called Walloon process, standard in seventeenth-century England. It meant melting down the sow or pig, decarburizing the iron by oxidation in the first hearth or "finery." Following another reheat, the now slaggy lump of pasty iron was forged under the large helve-hammer to remove the cinder and shape it roughly into the square cross section of the semi-finished stage.

After another reheat in the second hearth, the "chafery," the large rough bar was forged down into smaller bar stock, the principal sales items which a blacksmith could transform into an almost infinity of tools and implements. The Forge House has two water wheels – a large pitchback wheel for the hammer, and a smaller but undershot wheel for the bellows.

The last element of Hammersmith, the Slitting-Rolling Mill, is of great historic inter-

One of the reconstructed waterwheels. It is of practically all-wood construction even to the fastenings. The only iron fittings are the gudgeons or journals in the shaft ends and the shrunk-on hoops that prevent breakout.

est. It is a unique entity, since in all of contemporary Europe there is documentary evidence for only fifteen or so similar installations. This Mill absorbed perhaps twenty percent of the bar stock from the Forge House. The iron bar iron was cold-sheared into 2 or 3 ft (60 or 90 cm) lengths and then reheated (more bellows, more waterwheels). Passed through a pair of waterwheel-powered rollers while red-hot and plastic, the short bars were turned into long, flat, and thin strips about a quarter-inch thick and several inches wide. These strips were then quickly fed into the slitting unit in which a number of revolving steel-edged disks cut the wide strips into narrow strips of perhaps quarter-inch width called nail rods. On being put up into 60 lb (27 kg) bundles, they were ready for sale and reduction to nails.

It must be realized that the demand for nails for houses and barns was enormous. For the blacksmith to make nailer's rods by drawing bars down by hand was outrageously expensive. However, with the rod from the slitting mill, it was but necessary to point one end and head the other, which could be done at home in odd hours using the ever-present fireplace and simple tools. Making nails — finishing pieces of nail rod — was a cottage industry of no mean proportions in the colony as well as mother country.

The two restored 150-barrel stills, with a crude oil storage tank on the hillside above. Standard Oil Company's plaque marking the restoration in the 1930s is on the still to the right.

Pioneer Oil Refinery—1876

Newhall, California
Pioneer Oil Refinery Site, north of Los Angeles

The first deliberate pursuit of petroleum in America took place in Titusville of the Oil Creek area of northern Pennsylvania during the summer of 1859. Petroleum was known, having been found and used here and there for centuries. Tests made at Yale University suggested a number of products which could be distilled from the oil. Oil was now to be sought in quantity rather than collected from pools into which it had seeped. "Colonel" Drake, commissioned by the Seneca Oil Company of Connecticut and given a budget of $2000, hit oil at 69-1/2 ft (21 m), having had to invent the drilling technique. This first commercial well was so shallow that it needed pumping, the first flowing well not being drilled until 1861. Rock oil was said to have medicinal properties good for many ailments: three teaspoonsful three times a day for cholera morbus, liver complaint, rheumatism and consumption. It was a time when a new illuminant was urgently needed, for the supply of whale oil was decreasing, a situation that would turn critical with the Civil War soon to come. The organic oils used for lubrication were far from satisfactory, especially in the presence of the high temperatures encountered in steam engines striving for ever-higher pressures.

Oil came to be drilled for in California, too, possibly in 1866. Four successful wells have been traced to Pico Canyon not far from Los Angeles. To produce a salable product, a small refinery was built in 1873 by the Los Angeles Petroleum Company near Lyons Station, a stage stop about three miles north of San Fernando Pass, but the venture proved unrewarding.

Designated
September 27, 1975

Better drilling methods imported from the Pennsylvania oil fields improved the production of Pico No. 1, which, when deepened to 120 ft (36.5 m), produced 10 to 12 barrels a day. If today's 42 gal barrel was customary then, the production lay between 1600 and 1900 liters per day.

In 1876 the California Star Oil Works Company was organized and a 20 barrel still set up near Lyons Station. Pico No. 4 was worked over and driven to 370 ft (112 m) for an output of 25 barrels a day (3975 liters). Today, Pico No. 4 is still in production – at the rate of a barrel or so a day – and is the oldest working well in the west. When Pico No. 1 was pushed down to 175 ft (53 m) to produce 30 barrels daily (4770 liters), the need for a new refinery was manifest. Since Lyons Station had been bypassed by the just-built Southern Pacific Railroad, the new refinery came to Andrews Station on the outskirts of Newhall where it was close to the tracks.

The new refinery – the "Pioneer Refinery," whose construction was supervised by J. A. Scott, an experienced oil man – was completed in August of 1876. Storage tanks of 20 to 100 barrels (3200 to 15 900 l) were placed on a hillside above the stills, the oil flowing to them by gravity. Two of the stills, of 15 and 20 barrel (2400 and 3000 l) capacity, had been brought over from abandoned Lyons Station, and a new, 150 barrel (23 800 l) still installed. All were set on brick foundations and were direct-fired. The last still was duplicated, giving four in all. Fuel was the heavy residual oil from earlier refining runs, with steam injection for atomization and hotter flame. The condenser was a wooden box, 5 x 5 x 125 ft (1.5 x 1.5 x 38 m) containing approximately 1400 ft (425 m) of 2 and 3 in. (50 and 75 mm) diameter iron pipe submerged in water. The condensed oils then flowed by gravity to a lead-lined agitator for treatment with chemicals to improve their burning quality.

Several products came from this Andrews Station refinery. Benzene was made in small quantity; other things included a 300 °F (149 °C) fire-test safety illuminating oil for use on ships, railroads, factories, and mines; a light lubricating oil for general machine purposes; a heavy lubricant for saw mills, quartz mills and railroad journal boxes; and two grades of kerosene. This last was the breadwinner, even though several runs were necessary to get a salable product. Kerosene production averaged about 750 gal (2838 l) per day. The originally quite shallow wells were deepened to increase their output, No. 4

becoming a flowing well when finished off at 560 ft (170 m). More storage capacity was installed until the total was 3300 barrels (525 000 l).

An early run of high-grade kerosene was shipped to San Francisco in 1876 just as eastern kerosene started a steep price climb. The presence and further prospect of quality western oil quickened interest in California Star Oil Works Company to which investors were attracted. Oil lands were leased in the San Fernando district of Ventura County, and the beginning of an integrated industry for supplying oil to California and the bordering areas was laid, in large measure due to the efforts of D. G. Scofield, who in time became president of Standard Oil Company of California. Events moved quickly. A new and much larger refinery was built at Alameda in the Bay Area about 1880, allowing the older refinery in Ventura to be phased out by 1888.

In its short life the pioneer refinery at Andrews Station (Newhall) served the west with distinction. In the 1930 s the Standard Oil Company of California restored the two largest stills and opened the historic site to the public.

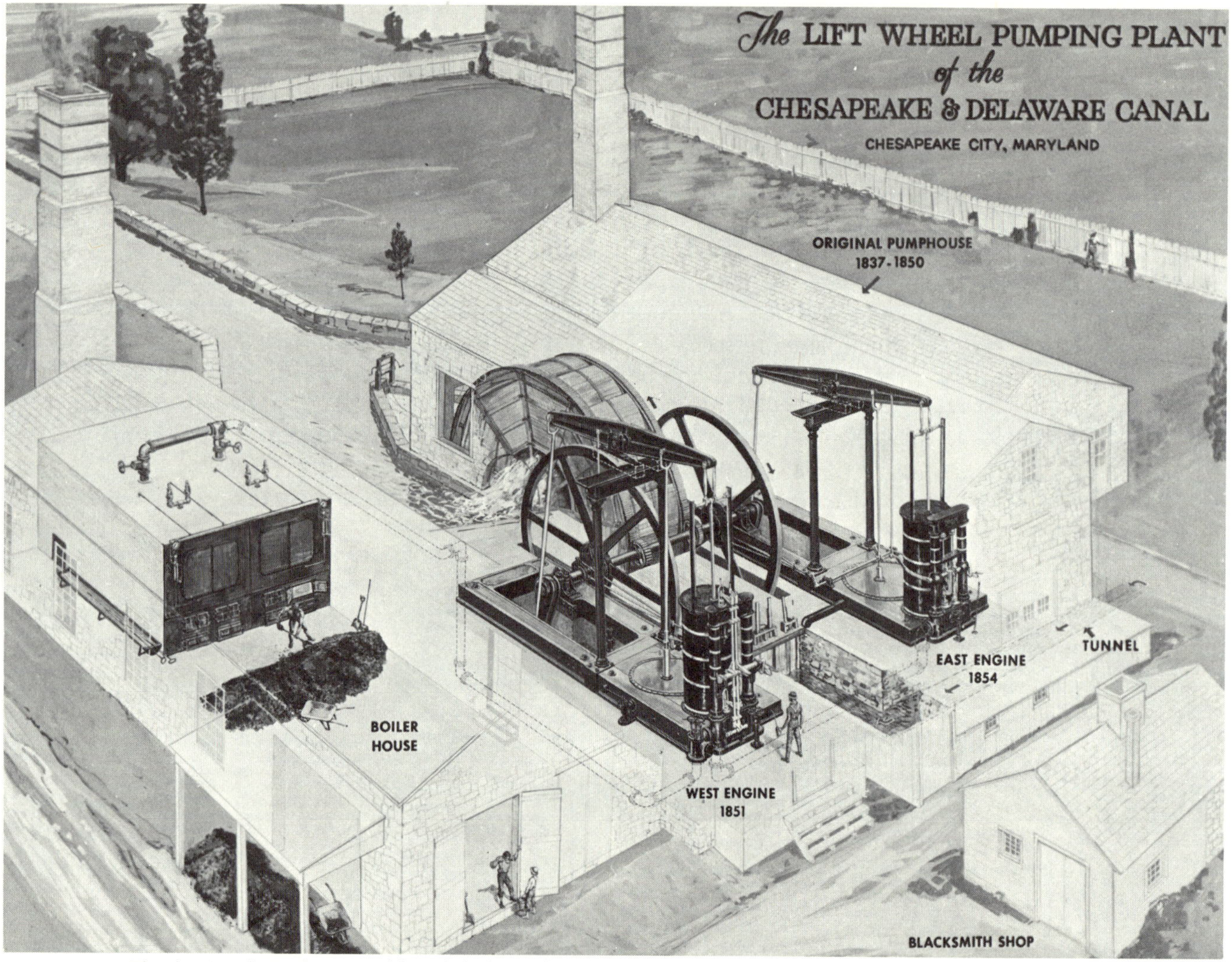

The drawing shows the unusual lateral discharge through the hollow hub of the 39 ft lift- or scoop-wheel. The rim drive — there is a ring or peripheral gear — allowed a lighter construction than if the power had been applied at the axle or hub.

Chesapeake and Delaware Canal Pumping Machinery—1852

Pump House Museum
Chesapeake City, Maryland

As canals go, the Chesapeake and Delaware is unique on several counts. It is short; even today, a century and a half after navigation began and many improvements, its length is not much different from the original 13-5/8 miles (22 km). The construction cost was very high because of terrain problems and generous dimensions. What is perhaps the most unusual aspect is that the canal – in highly modified form – is still in use, having become part of the coastal waterways system; most other canals ceased to function long ago.

The C & D Canal is a shortcut across the narrow neck of the 170 mile long (270 km) peninsula that is shouldered by Chesapeake Bay with Baltimore to the west, and Delaware Bay, the entry to up-river Philadelphia, to the east. The Canal shortens the Baltimore-Philadelphia water-distance some 280 miles (450 km) by eliminating the need of skirting the picturesque peninsula composed of pieces of three states – Maryland, Delaware and Virginia.

Earliest mention of the desirability of a canal linking the two great bays comes from 1661. A century or so later not only Philadelphia but also the State of Pennsylvania promoted a canal project that didn't receive favorable notice until 1799. Action came with the year 1800, when a $500,000 stock subscription was made. A Canal Company was incorporated in 1802, but it took years and the progressing construction of the Erie Canal (it would open in 1825) to provide the urge to get on with the local job in 1824. The C & D opened for navigation in 1829 and was an immediate success.

Designated
October 25, 1975

The Canal was one of the largest (not longest) of its day, for few canals anywhere exceeded its 66 ft (20 m) width at the surface and its 10 ft (3 m) depth. For example, the cross section of the vaunted Erie Canal [40 ft by only 4 ft (12.1 m by 1.2 m)] was but a quarter that of the C & D. More than 2500 men did the pick-and-shovel work of excavating in unfriendly territory that included a mile of tidal marsh and the famous Deep Cut, which, at the time, was regarded as being "one of the greatest works of human skill and ingenuity in the world." This cut, more than a mile long, was through the low ridge of solid rock forming the spine of the peninsula; at places it was 90 ft (27.3 m) deep. Then, too, there were earth-slides from the high banks, as when 375 000 cu yds (287 000 m^3) dropped into the canal. It is little wonder that the C & D cost-per-mile was high.

Lack of water at the summit is a common illness of canals. This disease, compounded by dry seasons, struck in the late 1840s. Tolls suffered because the "deep-draft" traffic had to be restricted. A steam pump of some sort was indicated, a machine capable of lifting 200 000 cu ft (5664 m^3) per hour a distance of 16 ft (4.85 m).

A prize contest, announced in 1848, brought forward fifty plans of which the design of Barnabas Bartols, engineer with Merrick and Sons, was selected, after a model and estimates of the construction and operating costs had been reviewed.

The water-lifting machinery went on-line in the spring of 1852. A single vertical-cylinder condensing beam engine of no-nonsense design having a 36 in. (920 mm) bore and 7 ft (2100 mm) stroke drove the flywheel shaft carrying two 4 ft (1219 mm) diameter pinions of 3.25 in. (82 mm) circular pitch and 11 in. (290 mm) face width. These pinions engaged two spur wheels of 39 ft (11.85 m) diameter forming the ends of a 10 ft (3.04 m) wide scoop wheel having twelve peripheral buckets or scoops. Water, scooped by the buckets when they were immersed, was discharged laterally through an opening concentric with the shaft. The engine's normal speed of 24 rpm turned the wheel 2.46 revolutions that delivered the contents of 29.5 buckets in a minute. The "dip" of the wheel was 20 in. (51 cm). The scoop wheel's simple design (going back to Roman days), while impairing the hydraulic efficiency, called for little maintenance.

A second Merrick engine, one with improved valve-gear to make for better economy, was added at the other end of the scoop wheel in 1854. The wheel's dip was increased to 28 in. (71 cm), and the engine speeds dropped to 18 rpm to give about 22 scoops per minute. An 1873 news article, "The Big Canal Wheel at Chesapeake City," told the readers that " – the immense wheel – is one of the curiosities of the country. There is none like it in the United States." If, as the article went on to say, 170 tons (154 t) of water were lifted their 16 ft in one minute, the "water horse-power" works out at 165. The article also spoke of two 150 hp (111 kW) engines "running 18 to 24 hours per day" consuming a daily 8 tons (7.25 t) of Cumberland coal. The pumping plant was in operation until 1927.

Wars have pointed up the usefulness of the Canal. When Virginia seceded from the Union in '61, Washington was left undefended by troops or fortifications; Federal forces were moved in via the canal. By the time these troops reached the capital, Confederate outposts were already threatening it. The lower Potomac was in Rebel hands and patrolled by a Confederate gunboat. Neutralization of the offensive measures by the C & D canal probably saved Washington from being captured. All through the Civil War the Canal provided a vital link for the movement of the Union's men and supplies, that, had they been forced to go around Cape Charles, would not only have exposed them to Confederate attack but also taken days longer for delivery. Similarly, in World War I, the Canal was an indispensable link in the movement of munitions and materiel because of avoidance of submarine attacks and the saving in time.

In 1919 the government purchased the Canal and turned its operation over to the U.S. Army Corps of Engineers. Much updating was done at this time, and during World War II the Canal once again became an important military adjunct. Today, as part of the Intracoastal Waterway system, the Canal is over 400 ft (120 m) wide and 30 ft (9.1 m) deep. Traffic on this, one of the world's busiest waterways, is monitored by radio.

The USS TEXAS on a trial run in 1913, quite different from her present appearance after the modernization of 1925. In particular, tripod masts replaced the basket or cage type, oil firing supplanted coal, and additional hull armor (torpedo blisters) increased the beam by 12 feet. U.S. BuShips photo.

Reciprocating Steam Engines USS TEXAS—1912

San Jacinto Battleground State Park
Houston, Texas

The steam engine was born, lived and died in the span of two centuries. Birth was difficult in 1712, and while there were hard times during maturation, the engines at the end of the era were indeed magnificent. Two very large engines of this time have been preserved in the USS TEXAS; they represent the ultimate in naval reciprocating engine construction, having reached the practical limits of size and horsepower. They are something of an anachronism, for they were built when the turbine was well-established for high-powered vessels such as liners and battleships.

When the time came to plan for building the USS TEXAS and three companion ships, the U.S. Navy had three turbine-driven battleships in service. A comparison of their performances with similar reciprocating-engined ships showed the reciprocators to be more economical at cruising speeds; the turbined USS UTAH used 20 percent more coal than the reciprocator USS DELAWARE at a speed of 12 knots. And there was another thing: it was accepted that most repairs to reciprocating engines could be made at sea using the ship's own facilities, whereas turbine problems were likely to be internal, as nozzle erosion, blade stripping and rotor corrosion, ailments requiring special dockyard facilities or even the turbine-maker's works. Then, too, the turbine manufacturers had a slow delivery record and had refused to abide by standards laid down by the Navy Department.

At any rate, the combination of greater cruising range and relative ease of upkeep of machinery seemed compelling considerations in 1910, since it was expected that the battleships might cruise for many months at a time thousands of miles from a home port. So it was reciprocating engines for the TEXAS, NEW YORK and OKLAHOMA, although the OKLAHOMA did end up with turbines after all.

Designated
December 1, 1975

The TEXAS' engines followed the standard design for express liners, high-speed channel steamers and warships, namely four-cylinder, triple-expansion with four cranks (at 90 deg) and two low-pressure cylinders, it being necessary to split the low-pressure into two cylinders to avoid excessive bores. As with most four-cylinder triple-expansion engines, the two low-pressure cylinders are at the ends of the engine, an arrangement which facilitated the balancing of the reciprocating forces. Vibration was a problem with reciprocators; in a large engine the reciprocating weight at each cylinder might go 4 tons (3.6 t), and at full speed the maximum piston acceleration could be of the order of 13 g. One of the reasons for the popularity of the four-crank engine for both warships and liners was the improved internal balance of those periodic inertial forces.

The TEXAS was a twin-screw ship, with over 14 000 hp (10430 kW) on each shaft. Each engine had a 48 in. (1220 mm) stroke; the cylinder bores were 39 in. (990 mm) for the high-pressure, 63 in. (1600 mm) for the intermediate-pressure, and 83 in. (2100 mm) for each of the low-pressure cylinders. A single low-pressure cylinder would have needed the monumental bore of 117 in. (9 ft-9 in. or 2971 mm)! With steam at 274 psi (1890 kPa), the ship had a speed of 21.05 knots (10.8 m/s or 39 km/h) with the shafts turning at 125 rpm, a speed appreciably higher than the 80 rpm that was about tops for merchant ships. No picture of an engine is available.

Steam was supplied by 14 Babcock and Wilcox watertube boilers having 62, 213 sq ft (5780 m^2) of heating surface over a grate area of 1554 sq ft (144 m^2). Coal fired, the closed fire-room carried a forced draft from a pressure of 2 in. (51 mm) of water. Some 2850 tons (2896 t) of coal were carried; conversion to oil-firing was made in 1925.

Manganese bronze was used for the built-up, three-bladed propellers having diameters of 18 ft, 7-1/14 in. (5.7 m) and a pitch of 19 ft, 11-1/2 in. (6.1 m).

The TEXAS was built at Newport News at a bid price of $5, 830, 000. The keel was laid in April 1911, with launching in May 1912, and commissioning in March 1914. The ship is 573 ft (174 m) long, has a beam of nearly 95 ft (29.9 m), a displacement of 27 000 tons (27 432 t), and carried a crew of over 1300. The ship's armament varied with time, since she was not decommissioned until 1948, after having seen service in two world wars. She was then presented to the State of Texas and recommissioned as the flagship of the Texas Navy and given permanent mooring in San Jacinto Battleground State Park. The TEXAS mounts ten 14 in. (355 mm) guns, sixteen 5 in. (127 mm) guns, eight 3 in. guns (76 mm), and 40 mm (1.57 in.) anti-aircraft guns. Three recoverable seaplanes were launched by a steam catapult on the stern. Belt armor of 12 in. (305 mm) protected the hull, the gun turrets had 14 in. (356 mm) armorplate, and 12 in. (305 mm) armor was used on the conning tower. A major conversion in 1925 provided a radical change in appearance, the old basket masts giving way to decked superstructures. Firing was changed to oil, and blisters added to the hull for torpedo protection increased the beam by 12 ft (3.65 m).

As an aside, the most powerful marine reciprocating engines belonged to two ships of the North German Lloyd. One ship, the KAISER WILHELM II won the Atlantic speed record in 1903. Each of the twin screws was driven by two engines coupled in line for a total of eight cylinders and six cranks. The engines were quadruple expansion, with the high-pressure cylinder over the first intermediate-pressure cylinder, the tandem pistons driving the middle cranks. Each engine developed 21 500 hp at 80 rpm; the 23 knot (12 m/s) speed record taken in 1903 was lost four years later to the Cunard Line's turbine-driven LUSITANIA and MAURETANIA. The German ships (there was a second and similar ship) were seized by the United States in 1917, and after brief service as troopers were laid up to be sent to the ship-breakers in 1940. Regrettably not even one of these unique engines was saved, whence the TEXAS' engines are the largest extant.

Turbines and generators in the Childs plant.

Childs-Irving Hydroelectric Project—1909/1916

Childs, Arizona
(Between Phoenix and Flagstaff on bad road from Strawberry)

Designated
April 10, 1976

It was just before the turn of this century that a cattleman, looking for sure range water for his herd, stumbled onto "the biggest spring he ever heard of" in the desolate but picturesque area about 70 air miles (113 km) north of Phoenix. The heavily mineralized water had given branches and stone a fossilized appearance, whence the water source became "Fossil Spring" and the runoff "Fossil Creek." That the Creek drops some 1600 ft (486 m) during the course of its ten-mile meander to the Verde River gave rise to thoughts about developing the springs for hydropower.

Hydroelectric power transmission, although recent, was a western innovation, the first system in the United States being a 13 mile (21 km) 3300-volt line between Willamette Falls and Portland, Oregon, built in 1891. Then came the 15 000 hp (11 200 kW) development at Niagara Falls in 1895 having terminal voltages of 2200. Mining projects in the copper-rich Irving region, and communities also, were looking for power. With a flow of 20 000 gal per minute (75 m^3 /min) proved by an engineering study, the Electric Operating Construction Company was formed in 1907 and construction of a generating plant at Childs was begun. This company was succeeded in the following year by the Arizona Power Company, the present operators.

Water from a dam below the spring was led to a small flat known as Dry Lake that could be used as a regulating reservoir for controlling the flow into the Childs penstock. This, now known as Stehr Lake, has an area of 28 ac (11.3 ha) and depth of 13.3 ft (4 m). The water course from spring to lake is a series of concrete flumes and tunnels approximately 7 mi (11.2 km) long. It provides a static head of 1075 ft (327 m) at

Childs for the three Pelton Wheel generators turning 400 rpm to produce in all 9000 hp (6700 kW) at 44 000 volts.

At the peak of construction, some 600 men were employed. Supplies were hauled by 150 wagons powered by 400 mules. It was necessary to build miles of road, switch-backed as it dealt with a sizable ridge; and a bridge was needed for the Verde River.

From Childs, a double-circuit 44 000 volt transmission line went to Prescott via Mayer and Poland Junction with taps to a number of mines. All three generators were on-line in 1909. The transmission line is unique for it is composed of windmill stands. It was impossible to transport wooden poles by mule, and steel towers as we know them had not been developed. However, pieces of windmill towers came in burro-sized packets, and were furnished by the U.S. Wind, Engine and Pump Company of Batavia, Illinois. The lower end of the penstock was forged by Krupp in Germany and sent to Los Angeles around Cape Horn, quite likely by sailing ship.

New copper smelters required additional power, a demand that was met by the Irving Plant building between 1914 and 1916. This was built between the spring and Childs, and has a head of 481 ft (146 m). It generates 2100 hp (1560 kW) in a single Allis-Chalmers Francis turbine at 900 rpm.

In the mid-1920 s, the growing city of Phoenix, with a population of 44 000, received 70 percent of its power from the Childs-Irving stations over a 75 mi (120 km) transmission line constructed in 1919. Both stations are still active with most of their original equipment, and although their region is served by new facilities such as a steam plant in Tempe with 33 times their generating capacity, they remain as living monuments to the past and originators of Arizona's first transmission system.

Hanford B Reactor after completion

Hanford B-Reactor—1943

Hanford Reservation, Washington
(Near Richland)

Designated
May 10, 1976

Plutonium, a chemical element of central importance to atomic weapons and nuclear engineering, was discovered in 1940. When the isotope Pu^{239} undergoes the fission reaction completely, it gives rise to a heat energy equivalent to 44.5×10^6 kW hr per Newton, which explains its effectiveness when used as an explosive ingredient for nuclear (atomic) weapons. For "peaceful" or industrial purposes, more or less conventional steam turbines convert the heat energy into convenient electrical form. The world became aware of the existence of plutonium with the dropping of the nuclear bomb over Nagasaki in August 1945.

Uranium-bearing ores carry plutonium in very small concentrations; the quantity production of the fissionable form, the isotope Pu^{239}, requires an enormous establishment focused on an extraction procedure carried out in devices known as nuclear reactors releasing vast quantities of heat. The chemical plants are massive structures designed to cope with the grave problems accompanying extremely high levels of radioactivity. Plutonium 239 is also one of the most highly radioactive pollutants.

Hanford Engineer Works – the prosaic name was chosen to preserve the secrecy of the wartime activity – became the location of the first large-scale production reactors, and was, when founded in 1943, an isolated area of about 670 sq miles (1735 sq km). Extensive studies of the region showed that both soil and sub-strata were capable of supporting enormous constructions. Fortuitously, plenty of aggregate was readily available for concrete, the most extensively needed material. Ultimately, some 50,000 personnel were engaged in the construction of the facility. Two dams, Bonneville and Coulee, were a sure source of electric power, with the Colorado River able to supply ample cooling water having little or no dissolved or suspended materials.

Originally, six sites lettered A through F, were chosen as potential locations for graphite-moderated, water-cooled nuclear reactors at Hanford. Of these, the one designated B became the first plutonium producer. It went to work in 1944, just two years after Enrico Fermi's first experiment at the University of Chicago. The scale-up was enormous: Fermi's "pile" had operated at 200 Watts, whereas the B-reactor was designed to function at 250 million Watts. With the exception of two years when idled, the reactor was in continuous operation until deactivated in 1968.

Hanford's B-reactor represented a major milestone along the road to a new form of energy. While many reactors have followed in more recent time, the Hanford-B led the way through being the first to be successfully operated. Aside from the physics of the situation, the research and the engineering and planning that reduced theory to practice to materialize safe reactor operation was a brilliant achievement truly without precedent.

Mechanical engineering entered into the design of the reactor, cooling system, shielding and auxiliary support systems. The reactor itself – the "pile" – required 2,000 tons (1968 t) of carefully machined graphite blocks laid to a tolerance of plus or minus 0.005 in. (0.127 mm) and bored to carry 2004 aluminum tubes hydrostatically tested to 350 psi (2411 kPa). There is an inner thermal shield composed of a 10 in. thick (25.4 cm) envelope of cast iron blocks, with the biological shield of Masonite and steel boosting the total weight to nearly 10,000 tons (9842 t). Supplementing this, room walls are 3 to 5 ft (0.9-1.5 m) thick with lead shielding affording additional operator safety. Elaborate control mechanisms regulate the pile activity; the operations are carried out entirely by remote control. Great quantities of cooling water, both raw and de-aerated, are circulated by an installed pumping capacity of 92,500 gal per minute (350 m^3/min) driven by either steam or electric power. Ventilation was handled by electrically- or steam-driven fans having a capacity of 100,000 cu ft per minute (2832 m^3/min).

The rotor of a centrifugal refrigeration unit of 140-ton capacity of the type installed at the 3600- to 4600-foot levels. Dr. Willis H. Carrier is the second from the right.

Underground Air Conditioning—Magma Copper Mine—1937

Superior, Arizona
(On US 60, east of Phoenix)

Designated
May 22, 1976

The Magma mine is a warm mine, the rock temperature at the 2000 ft (610 m) level being 109° F (43° C), with the temperature increasing at the rate of 1-1/2° F (0.83° C) for each additional 100 ft (30.5 m) of depth. In 1935 a vein at the 4000 ft (1220 m) level, with a rock temperature of 140° F (60° C), was to be developed. This was an impossible situation, and it was evident that air conditioning – cooling and humidity control – would need to be provided for levels below 3200 ft (975 m).

Current practice was to open up a level and let it stand for several years to drain and cool off from the air sent through it by ventilating fans placed on the level. The 3200 ft (975 m) level had required about three years of such passive cooling before it could be worked with any degree of efficiency.

The unique temperature problem persuaded the Magma management to call in Dr. Willis H. Carrier of the Carrier Corporation to make a survey and recommendation for what would be the first underground air conditioning system in North America. The earliest such installation anywhere had been made by Carrier in the Robinson Deep Mine in Johannesburg, South Africa, some years earlier. Almost two miles deep, the temperature was of the order of 130° F (54.4° C).

During the first half of 1937 two mechanical refrigerating units were placed at the 3600 ft (1100 m) level, the lowest developed level at the time. Since all equipment had to be lowered through 40 by 60 in. (1 X 1.5 m) shaft compartments, two 140 ton (137.8 t) units rather than a single larger one were selected. This choice gave flexibility, allowing one unit to be down for routine maintenance while the other carried the part load.

Power for each chiller unit was a 200 hp (149 kW), 2200 volt, 25 cycle three-phase induction motor turning 1440 rpm geared up to 6750 rpm for the compressor drive. After six weeks of operation, the temperature change on the 3400 ft (1036 m) level provided better conditions than those existing on the 3200 ft (975 m) level after three years of ventilating by fans without air-conditioning. In four months an effective working temperature of 71° F (21.6° C) has been reached at eight stations on the 3600 ft (1100 m) level.

A third machine was brought to the 4000 ft (1220 m) level in 1939; by the summer of 1941 a total of six machines with a combined capacity of 840 tons (826.7 t) had been installed. Since then, a number of other air-conditioning systems have been located as needed to provide comfortable working conditions in newly developed areas.

The Landmark is the original pair of 140 ton (137.8 t) refrigerating units of 1937 that dropped the temperature at the 4000 ft (1220 m) level from its normal and unworkable 140° F (60° C) to 93° F (33.9° C) in four weeks. This equipment was used until the surrounding area was mined out. Since it was judged that it would be too expensive to move the units, they were abandoned in place after being stripped of salvageable parts.

One of the original simple (single-expansion) engines of 1891 rebuilt as a compound engine, 1896. The 9- and 15-inch diameter pistons had a common crosshead with 22-inch stroke and drove to rocking beams from which connecting rods reached to cranks on the forward wheels. Both driven axles were fitted with cog wheels.

Pike's Peak Cog Railway—1891

Ruxton Avenue Depot
Manitou Springs (Colorado Springs), Colorado

A late-comer in the field – operation didn't begin until 1891, some 22 years after Mount Washington, the world's first – the cog railway to Pike's Peak reaches the highest elevation, 14 110 ft (4300 m), achieved by any rack railway, and its 8.9 mile length (14.3 km) is exceeded by only one, in Switzerland. Pike's Peak is not Colorado's highest mountain, but does have the good fortune to overlook the plains instead of being huddled among other peaks allowing no distant view.

Although the first surveys for the route date from 1870, progress was slow. An early company, organized in 1883, made a good start but collapsed following the failure of a New York bank that held the company's money. In 1888 a new firm bearing the present corporate name was organized, being financed entirely by Zalmon G. Simmons, among other things a manufacturer of mattresses in Kenosha, Wisconsin. Simmons, like the earlier Marsh on Mount Washington, had a rough time on his first trip up the Peak. Simmons or his heirs were the sole owners of the railway until 1925.

Passenger service to the peak began in 1891. Construction had been difficult. All grading had to be done by pick, shovel and wheelbarrow, the road being too narrow for the use of horse- and mule-equipment. Also, the "thin air" of the higher levels hampered the workmen. The line is 46 292 ft (14.1 km) in length, climbing from 6561 ft (2000 m) at Manitou Springs (just outside Colorado Springs) to the summit of 14 110 ft (4300 m), a rise of 7539 ft (2300 m). The average grade is 16.2 percent, the maximum 25. The standard gage road uses the Abt system of rack, patented in Switzerland in 1882.

Designated
May 24, 1976

Both Marsh, who built the earliest cog railway up Mount Washington in 1869, and Riggenbach, who built the world's second in Switzerland in 1871, used racks in which the "teeth" were bolts connecting the upright flanges of angle irons, giving an effective and ladder-like rack. The Abt design features conventional gear-tooth racks, two or three racks being laid alongside each other with staggered teeth. On the Pike's Peak road only two racks are used, the pitch of the teeth being 2.35 in. (60 mm). The thickness of the steel — the width of the rack — depends upon the grade, 7/8 in. (22 mm) being used for grades up to 12.5 percent while going to 1.25 in. (30 mm) for the steepest grade.

The original motive power consisted of three 26 ton (23.6 t) Baldwin tank-locomotives, with the bearing frames inclined at 9 degrees so that the boiler was level on the 16 percent average grade. The bore and stroke of the two-cylinder engine were 17 and 20 in. (430 and 510 mm), with three double pinions or cogwheels for the drive. A fourth Baldwin engine, a Vauclain compound with cylinder bores of 9 and 15 in. (230 and 380 mm) and 22 in. (560 mm) stroke, delivered in 1893, was such a decided improvement over the three simple engines that the older engines were sent back to Philadelphia for re-design. Other steam engines were added as necessary. Train control on the descent involved having the engine compress air, with a steam brake as well. As on other cog lines, the engine nosed the car up the hill — there were no couplers.

With 1925, The Pike's Peak Auto Highway Company acquired the road from the Simmons family and modernized the stable of motive power. A revolution in railway passenger service in the United States took place when the Burlington's "Zephyr," the first streamlined, light-weight diesel-electric train went into service in November 1934. In this spirit the Company built a 24-passenger railcar in 1936, a "streamliner," in its own shops, using a 175 hp (130 kW) gasoline engine before turning to a Cadillac engine. The internal-combustion engine experience was encouraging and led to the decision to phase out the steam power in favor of diesel-electric. The first diesel-electric (built by General Electric and powered by General Motors) went into service in 1939, and brought with it electrodynamic braking for the down-trip, the engines being off.

Steam's last revenue trip was made in September 1958. It is rewarding to find a far-seeing Company management that has spared four of the oldest steam engines from the wrecker's torch. They are on display at end-of-track, Manitou Springs Station, Cheyenne Mountain Zoo in Colorado Springs, and the Colorado Railway Museum in Golden.

One of two Swiss-made railcars added in May of 1976. Each 1200-horsepower, 122-foot-long train can carry 206 passengers. The double-rack cograil is visible between the running rails.

Since General Electric was really not interested in building one-off units such as railcars, Swiss equipment with Cummins diesel engines was obtained, the first being delivered in 1963. With the 1970 s two twin-unit diesel-hydraulic trains seating 206 passengers (the earliest steam trains carried only 50 passengers) were ordered from the Swiss Locomotive Works of Winterthur. They went into regular service with the designation of the Cog Railway as a National Historic Mechanical Engineering Landmark in May, 1976, and are the first articulated trains on any rack road. Each car has two American-made diesel-hydraulic systems of 300 hp (223 kW) for a total of 1200 hp (890 kW) per train.

The initial installation on the banks of the Fore River, 1925. Although the now-expanded station is oil-fired, the slanting coal chute spanning the access street to the Fore River Bridge remains.

Edgar Steam-Electric Station—1925

Weymouth, Massachusetts

In the early 1920 s, when steam pressures of the order of 300 psi (2067 kPa) were common to the electric utility industry, a 500 psi (3445 kPa) steam plant was reportedly under design. New materials such as chrome-molybdenum were being recognized for properties suiting them to the temperatures of really high-pressure steam that would provide substantial gains in efficiency. It was in Boston, on the occasion of providing a necessary new station, that a very bold step was taken by going to 1200 psi (8300 kPa).

The visionary responsible for the major step was Irving E. Moultrop (1865–1957) of the Edison Electric Illuminating Company of Boston who had long been active in the work of the ASME Boiler Code Committee and would become Chief Engineer of the Company. Proceeding with caution, it was decided to include for the Edgar Steam-Electric Station at Weymouth one 1200 psi (8300 kPa) boiler and turbine working with a more conventional 350 psi (2400 kPa) steam-turbine system, the latter in effect operating on the exhaust steam of the high-pressure unit. The arrangement as worked out had 1200 psi (8300 kPa) steam at 700 °F (370 °C) expand down to 375 psi (2600 kPa) and 500 °F (260 °C) in a 3150 kW General Electric turbine. This exhaust steam was reheated to 700 °F (370 °C) and discharged into a header already supplied with 350 psi steam at 700 °F feeding two "normal" turbines each rated at 32 000 kW that expanded the steam down to one inch Hg (25 mm) absolute. The station was designed and built by Stone & Webster of Boston under Mr. Moultrop's direction.

Designated
May 27, 1976

The 1200 psi boiler was a modified Babcock and Wilcox cross-drum type boiler of conventional design. The drum was a 32 ft (9753 mm) long steel forging of 4 ft

CHARLES LEAVITT EDGAR STATION OF THE EDISON ELECTRIC ILLUMINATING COMPANY OF BOSTON

The 1200-psi turbine of the "Edison Electric Illuminating Company of Boston."

(1220 mm) diameter and 4 in. (101 mm) wall thickness made in the gun works of the Midvale Steel Company.

A new record for economy was established when the first phase of construction went into commercial operation at the end of 1925: a kilowatt-hour of electricity was being produced from less than 1 lb (0.45 kg) of coal when contemporary "electric light" plants required 5-10 lb (2.2-4.4 kg) for the task. The performance of the first highest-pressure unit was observed by utilities the world over and became their model for expansion. The now-standard X-raying procedures for piping and turbine casing stem from this world's first 1200 psi (8300 kPa) installation.

With two years experience in hand, a first extension of Edgar Station (named for Charles L. Edgar, president of Boston Edison for 32 years) saw two 1400 psi (9400 kPa) boilers feeding a 10 000 kW turbine whose exhaust steam would be turned into the 350 psi "normal pressure" line for the 65 000 kW main generating unit.

Edgar Station has moved to the position of an auxiliary, its one-time forefront equipment having been overtaken by the technology it pioneered. An addition to the plant in 1947 featured a turbo-generator giving a 30 percent increase in economy. In the course of events the boiler firing changed, going from coal to pulverized coal and oil, which, more recently, was modified to allow the use of natural gas as an alternative fuel. With the removal of the earliest equipment following retirement in 1971, the plant operation is winding down, although it is currently scheduled to be kept in operation until 1991.

The operating model (with trunnion-supported boiler) used by Marsh on the New Hampshire legislature in 1858 when requesting the charter for the cog railway.

Mount Washington Cog Railway—1869

Railway Base Station
Mount Washington, New Hampshire

This joint ASME-ASCE Landmark is unique on two counts. First, it is, for all practical purposes, the pioneer cog railway of the world; and secondly, it is still in primeval condition, not having been modernized by the substitution of electric or diesel power for steam engines that are still coal-fired. All the excitement, noise, smell and cinders of the typical steam operation of 1869 are still to be experienced, a rare adventure.

Actually, the first cog or rack locomotive dates from 1812, when the English colliery owner Blenkinsop, in doubt about wheel traction for performance on the level, had a rack-drive locomotive constructed, a precaution proving unnecessary for ordinary railroading. The mountain rack locomotive was first used on Mount Washington in 1869, to be followed by the Vitznau-Rigi Bahn in Switzerland in 1871. The Pike's Peak operation dates from 1891.

Versatile and imaginative Sylvester Marsh (1803–1884), a business man and engineer born in New Hampshire who made a fortune in Chicago, was caught in a bad storm while climbing Mount Washington in 1852, and resolved to build a railway up the mountain that is the highest peak in the northeastern United States. The weather at the treeless summit is often called the worst in the world – the buildings are anchored, and windspeeds of 225 mph (360 km/h) have been measured at the Observatory.

Marsh retired in 1855 – he was 52 years old – to work up plans that he presented to the New Hampshire Legislature in 1858. Part of his arguments were made with the aid of a model now held by the New Hampshire Historical Society. He received the necessary charter in 1858, but poor business conditions delayed the beginning of construction to 1866.

Designated
June 26, 1976

All material was freighted to the Base Station site (present-day Marshfield) by ox-train from Littleton, 25 miles (40 km) away. A water-powered sawmill was set up for getting out the timber — not only for cross ties, but also for the 3 miles (4.8 km) of the 3½ mile (5.6 km) track that had to be built on trestles because of the ugly terrain. The line was laid out with a gage of 4 ft-8 in. (1422 mm).

The ladder-like between-rail rack was made of two angle-irons with 1.5 in. (38 mm) diameter bolts on 4 in. (102 mm) centers serving as rack teeth. It would be interesting to know whether the 111 000 holes were drilled or punched. Marsh designed the first locomotive that was constructed by the Massachusetts firm of Campbell, Whittier and Company in Roxbury. The pieces were brought to the Base Station in 1866 and reassembled to give Locomotive No. 1, "Hero." Its upright boiler, swung on trunnions that it might stay vertical on the grades, made the engine look a bit like the pepper-sauce bottles of the day — hence the popular name "Peppersass." Two cylinders powered the front axle through gears, a single cogwheel working the rack for the stately speed of 2 mph (3.2 km/h). The cab was open, and there was no tender. A ratchet-safety cog was provided to check roll-back. On the descent the pistons compressed air to retard motion and supplement the hand brakes. The cars were pushed up — no couplers needed.

"Peppersass" was retired in 1878, but went to the Columbian Exposition of '93 in Chicago and to the 1904 Louisiana Purhcase Exposition in St. Louis. Somehow the old engine was then lost, but on being found in 1920, was returned to New Hampshire to be overhauled and slicked up by the Boston and Maine Railroad. Big plans were made for the return of "Peppersass" to "her mountain"; the big day was July 29, 1929. Governors from six states were on hand, as well as 500 official guests, rail fans, news photographers, correspondents and even sound equipment (talkies had just come in). To shorten a long story, "Old Pep" was steamed and given the throttle. She lost an axle due to fatigue failure and teeth broke on the gear wheels, allowing the engine to rush down the line and shoot off the track. Of the several people aboard the engine, one was killed. The engine was rather well scattered around, but was reassembled by the end of the summer and put on permanent display at the Base Station.

The railway, opened to traffic in July 1869, starts partway up the mountain at an

Locomotive No. 2, "The George Stephenson."

Train on Jacob's Ladder section with grade of 37.4 percent (20.5 degrees).

elevation of 2688 ft (819 m). It climbs 3600 ft (1100 m) to the summit of 6288 ft (1916 m): this summit elevation is roughly that of Manitou Springs where the Pike's Peak Cog Railways *starts* its climb to 14 110 ft (4300 m). The 3.5 mile (5.6 km) railway has an average grade of 25 percent (14 deg), with a maximum of 37.4 (20.5 deg), this steepest portion being known as Jacob's Ladder, an allusion to the Biblical ladder ascending into heaven.

New Hampshire manufacturers, as Walter Aiken of Franklin and the Manchester Locomotive Works, have furnished several of the locomotives, "horizontal" boilers being introduced in 1878. Currently representative motive power is an 0-2-2-0 four-cylinder engine whose odd appearance fits it to the track. The balloon-stacked boiler, slanting at 10 degrees to keep water over the crown-sheet, is followed by a huge cab; the locomotive gives the impression of kneeling. A spidery-wheeled tender trails the locomotive that shoves the single car ahead of itself – there is no coupler. The journey is "slow, cold, cindery, noisy, disarming" (in the words of David Morgan); the locomotive with its four cylinders has the impressive sound of a heavy-traffic Mallet pounding through the Rockies. The speed varies between 2.5 and 1.75 mph (4 and 2.8 km/h) on the 70-minute uphill fight. The engines are gutty little machines that present the sound and sight and smell of that vanishing thing, a steam operation.

Marsh seems to have had a lot of fun from his hobby that he said cured his "dyspepsia"; part of the pleasure may have been due to the eventual financial success, for the first dividend, paid in 1877, was nine percent. By the time of Marsh's death in 1884 investors had received an 88 percent return. It was now that the Boston and Maine Railway acquired the line, operating it until 1930 when it was sold to Henry N. Teague who managed it successfully despite the depression years. Teague deeded the mountain property to Dartmouth College that in 1962 sold the line to Col. Arthur Teague (no relation). Col. Teague saw to a good deal of redesign and made improvements during the years before his death in 1967. Since then, Mrs. Teague and capable associates have been running the Railway with enthusiasm and devotion to keep it one of the splendid tourist attractions of the northeast.

Four of the six generators that were reported to be the largest three-phase machines of the time, 1895. Their combined capacity was 3000 kW.

Folsom Powerhouse No. 1—1895

Folsom, California
(North-east of Sacramento)

The Old Folsom Powerhouse on the American River just below Folsom still shelters the machinery that in 1895 generated 3000 kW for the streetcars and other commercial and private users of electricity in downstream Sacramento, 22 miles (35 km) away. Although not the only nor the first hydroelectric plant in the country, its transmission line was three times as long as the better known contemporary installation between Niagara and Buffalo. This "longest" line of the time, carrying three-phase power at 11 000 volts did much to prove to the skeptical that electric current could be sent over considerable distances.

The story of Folsom begins with gold rush days. Not all of the adventurers of the time sought gold, some seeing their fortunes in other resources of the "new" country, as in the magnificent timber stands, or the exploitation of water power for which they envisioned industrial complexes such as those in New England, taking Lowell as an example. One emigré from the east, Horatio Gates Livermore of Maine, together with his sons, acquired land and also the water rights to the American River. A dam at Folsom would not only provide a holding area for logs, but also furnish water for factories and create a storage pond for the irrigation of orchards and vineyards. The dam was finished in 1893, having been five years in the building, much of it done by convict labor.

Designated
September 12, 1976

By the late 1880 s it was becoming apparent that water power might be better used if converted to electricity that could be shipped anywhere by wire than if made a direct motive force. Transmission-line experience was building, not only in this country but in Germany and Italy as well, and the idea of sending power to established Sacramento came into ascendancy. Progressive Sacramento already had a small electric railway system oper-

The original twelve air-blast transformers, later replaced by Westinghouse units.

ating from batteries. Competition came with the formation of the Sacramento Electric Power and Light Company that obtained a street railway franchise and soon acquired the old transit company by merger. Electricity came to Sacramento along with a bang on July 13, 1895, a 100-gun salute marking the event at four o'clock in the morning as the first current came down the 22-mile-long (35 km) transmission line from Folsom.

The power originated with four pairs of 30 in. (760 mm) McCormick turbines under a head of 55 ft (18 m) capable of developing 1260 hp (938 kW) each at 300 rpm directly coupled to four generators with a total output of 3000 kW. These machines, built by General Electric and shipped around Cape Horn, were the largest three-phase generators of their time. They stand almost 9 ft (2.9 m) high and weigh 29 tons (26 t). This equipment, and the marble-faced control panel are still intact, although taken out of service in 1952 after 57 years of continuous duty.

An additional powerhouse installed in 1897 to generate some 750 kW is of interest because of the drive between turbine and generator that were separated because of possible flooding. The mechanical connection between the two was effected by means of a continuous or single hemp-rope drive. Some 2200 ft (720 m) of hemp in one strand were wound around the several grooves of the driving and driven pulleys, the rope going from the last groove back to the first, with the whole system kept taut by means of an idler pulley on a tension carriage. This continuous or American system could handle vertical drives, an impossibility with the English or multiple-rope system in which there were a number of parallel ropes, one in each groove, the driving tension having to be derived from the weight of the rope.

Transporter with mobile launcher and SATURN V.
Note the figures of men.

Crawler Transporters of Launch Complex 39—1962

Kennedy Space Center, Florida

Designated
February 3, 1977

These Landmarks – they are a pair – have the distinction, so far unique, of being recognized in their day, at a youthful age when they are going about their tasks with the quiet assurance of confidence and competence. The Landmark designation recognizes the engineering audacity that created the world's two largest land vehicles. Each transporter weighs about 3000 tons (2720 t) and can carry a load the double of that, such as the 6000 ton (5440 t) Saturn V rocket and mobile launcher combination, while traveling the several miles between points of assembly and launch. Space shuttle vehicles (or even others) can also be handled when fitted with proper adaptors.

The brief history of the transporters begins in 1961 with President Kennedy's goal of making a manned landing on the moon before the end of of the decade. The conversion of the goal into fact became the problem of the National Aeronautics and Space Administration (NASA). The task was divided, of course, with the Kennedy Space Center (KSC) receiving the responsibility for the designing of the launch complex for the Saturn V launch vehicle and Apollo spacecraft. What with one thing and another, it was decided that the Saturn V would be assembled and checked out under the shelter of the Vehicle Assembly Building (VAB), and then be somehow transported to a launchpad some miles away. The rocket, launcher and portable service structure would weigh about 6000 tons (5440 t) and need to be moved from 3.5 to 5 miles (5.6 to 8 km), the distance dependant upon the choice of the launchpad.

Three transport-concepts were considered – a barge and canal system, a rail system, and a land transporter. The last mode was selected as the most promising, and in course, a contract for the construction of two crawler-type transporters was awarded the Marion

Power Shovel Company of Ohio. This organization had many years of experience in the construction of large, self-propelled strip-mining power shovels, the largest having a 48 ft (15.7 m) square chassis. The completely self-contained KSC transporters are 131 ft long and 114 ft wide (40 by 35 m), an area 6.5 times that of the roughly similar mining equipment. Both transporters came into service in 1967. Unloaded they can wheel along at 2 mph (1.6 km/h); their loaded speed is half that. The two cost $14 million.

Each transporter has four trucks about the size of a large Greyhound bus. A truck has two tracks 7 ft, 6 in. wide (2.3 m) and 41 ft, 3 in. long (12.6 m), consisting of 57 treads or lags weighing one ton (0.9 t) each.

On-board diesel engines totaling 7630 hp (5684 kW) supply electric and hydraulic power. The trucks mount four traction motors each – there are then 16 in all – that are fed by four 1000 kW dc generators driven by two 2750 hp (2048 kW) engines. Two engines of 1065 hp drive two 750 kW ac generators for leveling, jacking and steering systems. Five thousand gallons (18 900 l) of fuel are carried; the fuel "economy" is about 150 gal per mile (335 l/km).

When on the road, the transporter is maintained level for two reasons. Obviously the launch assembly should be kept vertical – the maximum deviation is held to plus-or-minus 10 minutes of arc, or about a foot in the 363 ft (110 m) launch vehicle height. Also, the ramp to the top of the launch pad has a 5 percent grade that must be recognized. Secondly, a level vehicle means that the structural loads generated by the statically indeterminate four-point support are properly distributed.

The heart of the leveling system consists of two horizontal, mercury-filled manometer tubes 135 ft (41 m) long that cross diagonally under the deck to form an X (the principle is that of the 2000-year-old Roman water level). Transducers sense the errors and actuate pumps for hydraulic cylinders at each of the load-supporting corners of the transporter.

Each end of the transporter has a glass-enclosed control cab housing an operator and observer on the "right-hand" corner of the direction of travel. Since rearward and offside visibilities are limited, three outside observers maintain radio contact with the "driver." The highly trained crew of engineers and technicians numbers eleven.

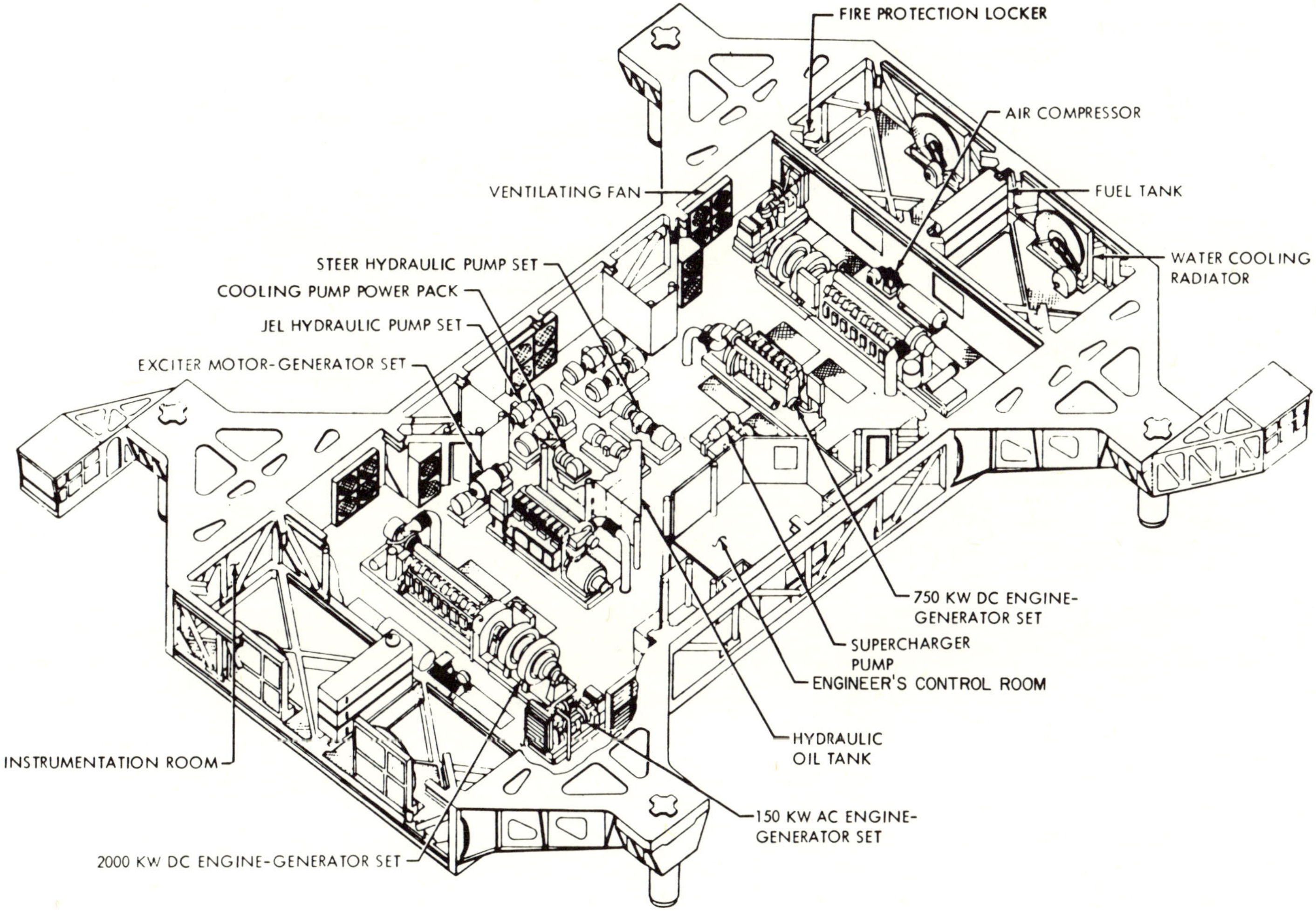

Equipment location on a crawler transporter.

The crawler transporter, 131 by 114 feet overall. Each of the four crawler trucks is about the size of a Greyhound bus.

After rocket assembly and checkout of the mobile launcher in the VAB, the transporter is driven under it and accurately positioned. Sixteen hydraulic jacks — four at each corner — are then activated to raise and so bring the transporter into contact with the overhead mobile launcher, lifting the latter clear of its ground-support pedestals. The run to the launch pad is now made. After arrival at the pad and proper positioning and so on, the transporter disengages itself and retreats to a safe distance to wait out the launch, after which the operations are reversed and the mobile launcher is returned to the VAB.

Sketch from the river-driven waterwheel period that began in 1822 and terminated in 1866, although phase-out had begun in 1851 as Jonval turbines were substituted. The wheels, turning at about 12 rpm, had diameters of 15 to 18 feet and 15-foot widths and drove double-acting pumps.

Fairmount Water Works-1815/1911

Fairmount Park
Philadelphia, Pennsylvania

Although sixteen American cities had municipal waterworks of some sort before 1800, the largest of them all, Philadelphia, was not among them. It was in 1799 that Philadelphia's City Council decided to lay water pipes in the principal streets for householders' convenience and fire protection. After much discussion it was decided that two rotative steam engines on the "Watt plan" – low pressure steam and condensing – would handle the water problem in two successive lifts. The first engine on the Schuylkill at the foot of Chestnut Street would raise water about 50 ft (15.2 m) to a tunnel leading to the second engine at Centre Square (now Broad and Market Streets), where the water would be lifted a final 36 ft (11 m) to a 16 000 gal (74 m^3) tank on the engine house from where the city distribution would begin. (These were ale gallons of 282 $in.^3$ [4.62 l], just a shade larger than the present Imperial gallon of 277.4 $in.^3$. The present U.S. gallon has 231 $in.^3$.)

The engines and pumps were built in New Jersey. Both engines had a stroke of 6 ft (1820 mm); the bore of the first was 40 in. (1020 mm), that of the second, 31 in. (810 mm). With 963 000 gallons (4500 m^3) being pumped in 24 hours, the "water horsepower" comes out about 18; what the rated engine powers were does not seem to be matter of record. If Philadelphia's population was 42 000 in the year 1800, the pumpage would have given each person about 23 gallons (102 l) per day, a generous figure since London is said to have furnished only a daily eight gallons per person in 1808.

Designated
March 29, 1977

The engines, considered to be the highest achievement of American manufacture, did well for ten years, but there was a problem with the boilers. Originally of wood with internal iron flues, they were 9 x 9 by 14 ft long (2.73 x 2.73 x 4.25 m) and made of

5 in. (127 mm) white pine planks held together by long staybolts and oak framing. Difficulty in maintaining steam-tightness, even though the pressure was not more than 2 psi (13.8 kPa), led to replacement boilers made of cast iron.

By 1811 the growing city's strain on the water supply prompted the Watering Board to establish a new steam pumping station and reservoir on Fairmount, where five acres were purchased in 1812 for the waterworks. A building was erected and contracts let for two engines. One was a low-pressure, condensing beam engine of 44 in. (1120 mm) bore and 6 ft (1800 mm) stroke with a 24 ft (7300 mm) beam of cast iron – a Boulton and Watt type – with cast iron boiler to furnish steam at 2.5 psi (17.2 kPa) gage pressure.

The contract for the second engine was given Oliver Evans, an already distinguished machine and engine builder who, with Trevithick in England, was a pioneer high pressure advocate, where high pressure meant anything over say 30 psi (207 kPa) gage. Evans' engine, his Columbian type also called "grasshopper," seems to have had a 20 in. (508 mm) bore and 5 ft (1520 mm) stroke, running at about 25 strokes per minute on steam that probably averaged 150 psi (1033 kPa) but went to 220 psi (1515 kPa) on occasion. There is a suggestion that it was rated at 100 hp (74.6 kW), making it the most powerful engine in America. On the basis of pumping 3 556 000 gallons (16 180 m^3) to a height of 98 ft (30 m) in a day when working at 24.75 strokes per minute, the "water horsepower" reckons to be 73.5 (55 kW).

Both engines worked side-by-side during the period 1815-1822, the high-pressure boilers exploding twice during this period, accounting for at least three deaths. Disappointingly, the operating costs of the engines were half again as much as the derived revenue. The performance of the Evans' engine seems to have been marginally the better. What with the hazard and the expense, a decision of 1820 proposed the substitution of river-driven waterwheels for steam power. The river was dammed and a 420 ft (128 m) long millrace brought water to breast-wheels driving pumps in the new Classical Greek Revival building that is today's reminder to the past glory.

The steam engines were decommissioned in 1822, three waterwheels (no fuel cost!) picking up the burden then. Within a few years the total plant was considered one of the engineering wonders of America. By 1842 the demands of the growing city had eight

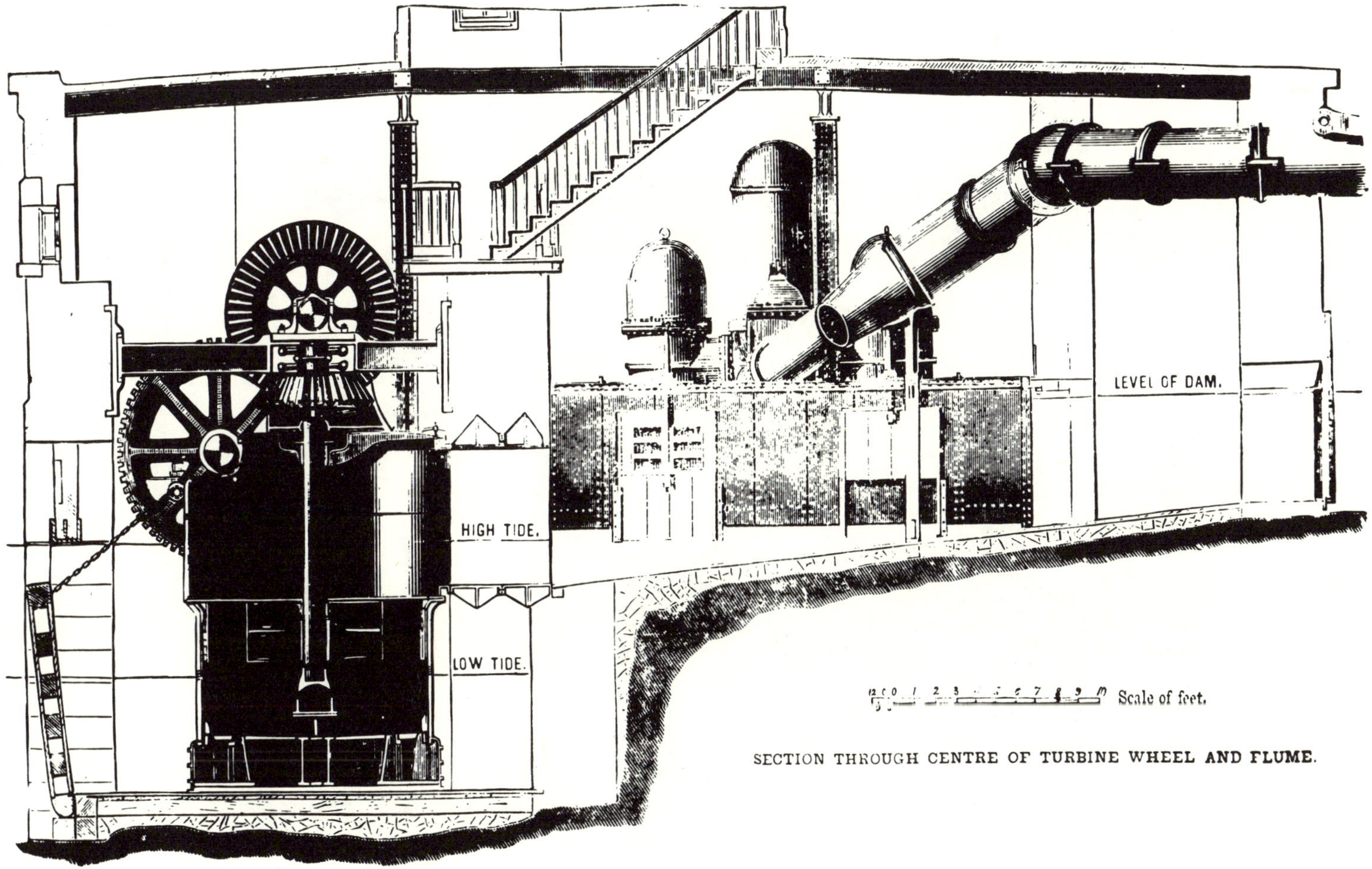

Schematic from the Jonval turbine period that had its beginning in 1851 as the waterwheels were replaced, and that came to an end in 1911 when an entirely new system with filtration plants took over.

wheels and four hill-top reservoirs in service. The wheels were of 15 to 18 ft (4.5 to 5.5 m) diameter but all of the same width, 15 ft (4.5m). They turned at 11 to 14 rpm, driving 16 in. (406 mm) double-acting pumps having 4.5 to 6 ft (1370 to 1830 mm) strokes.

These wheels were in service until 1866, having been gradually phased out from 1851 onward by Jonval turbines. These were vertical-shaft, outward-flow machines derived from the Fourneyron type suited to low heads. These turbines continued to supply portions of the city until March 1911, when new filtration plants took over all service.

The Fairmount Water Works also bear the distinction of being a National Historic Civil Engineering Landmark (June 1975).

Model of 100 hp (75 kW) Evans "Columbian" type Water Works engine, 1818, on exhibit in the Franklin Institute. Philadelphia's second steam pumping system of 1815 had two quite different engine designs. One was a traditional "Boulton and Watt" type low-pressure, condensing beam engine. Its mate was an Evans "Columbian," a high-pressure, non-condensing engine with a grasshopper parallel motion. Dr. Thomas P. Jones of the American Philosophical Society had operative live-steam models of each constructed for lectures held in 1818 that gave a "thrilling demonstration of the comparable merits of low and high pressure steam by means of two models, ..." of the new Water Works engines.

The model is about 42 inches (1070 mm) long. The bore and stroke are 2-1/8 and 6 inches (54 and 152 mm). Linkage details are shown very nicely, but the pumping arrangement seems proportioned to a table-top model rather than simulating the actual pump. The dimensions and fate of the other model are not known, nor are the conclusions that may have been reached at the end of a lecture.

Since no full-scale "Columbian" engine seems to have been preserved (the last known was scrapped following a fire in 1906), this large model from 1818 is a unique link to the past.

Back view, from the low-pressure end, of the 8425 ihp three-cylinder triple-expansion port engine (the USS TEXAS has a four-cylinder triple expansion engine, dividing the lowest pressure between two cylinders of equal size). Built-up wrought iron columns support the cylinders and crosshead guides. The reversing shaft is in the foreground. A Stephenson link motion actuates the piston valves.

Vertical Reciprocating Steam Engines–USS OLYMPIA—1891

Philadelphia Navy Yard
Philadelphia, Pennsylvania

Designated
March 30, 1977

The USS OLYMPIA is best known as Admiral Dewey's flagship at the Battle of Manila Bay, The Philippines, where, in May 1898, the entire Spanish squadron was destroyed in the first action of the Spanish-American war. Spain ceded the Philippines to the United States in December 1898 for 20 million dollars; and the islands were proclaimed an independent republic by the President of the United States on July 4, 1946.

The OLYMPIA is otherwise distinguished as well, for she was one of the first naval vessels to be fitted with vertical reciprocating engines, a departure from the usual practice of having propulsion machinery with horizontal cylinders to give a low profile below the waterline and thus be less vulnerable to the effects of gunfire. Merchant vessels had taken up the vertical engines earlier as the screw propeller displaced the (side) paddle steamer.

As part of the new American steel-and-steam Navy, the construction of the OLYMPIA was authorized in 1888 as Cruiser Number 6. To establish complete domestic capability for naval construction, the contract was awarded to Union Iron Works of San Francisco. The West Coast location meant procurement difficulties of armor plate, gun-forgings, machinery and so on that delayed construction: the ship was not launched until November 1892; commissioning was not until February 1895.

At this time there were two cruiser categories, the protected cruiser such as the OLYMPIA and the armored cruiser. Protected cruisers were of moderate size with a large number of medium-caliber rapid-fire guns and a curved protective plate of armor over the ship's vitals just above the waterline. By contrast, the armored cruiser was a larger, more agile ship, having the additional protection of side armor and carrying 8 or 10 in. (200 or 254 mm) guns.

The OLYMPIA has twin screws of 14-3/4 ft (4.5 m) diameter each of which was driven by a three-cylinder triple-expansion engine of 8425 hp (6276 kW) at 139 rpm with steam at 160 psi (1100 kPa) for a speed of 21.6 knots (11.1 m/s or 40 km/h). A stroke of 42 in. (1070 mm) was common to all cylinders, the bores being 42, 59 and 92 in. (1070, 1500 and 2340 mm). Four double-ended and two single-ended Scotch boilers with a total of 40 furnaces under forced draft on a closed stoke-hold system supplied the steam.

The OLYMPIA was of modest proportions, having a length of 340 ft (104 m), a breadth of 53 ft (16.2 m) and a displacement of 5870 tons (5963 t). Her armament was upgraded as the service demanded, but included four 8 in. (203 mm) 35-caliber guns among others, and six 18 in. (457 mm) Whitehead above-surface torpedoes. The protective armor was 4-3/4 in. (120 mm) on the slopes, but reduced to 2 in. plate (510 mm) on the flats. Trust in steam power was not absolute, since the ship carried also the auxiliary sail-rig of a two-masted schooner. She had a complement of 428.

Originally the flagship of the Asiatic Squadron, the OLYMPIA became the flagship of the small Caribbean Division in the 1900 s to spend her time in the North Atlantic. In World War I the ship stood North Atlantic patrol duty, shifting to the Mediterranean, Adriatic and Black Seas at the end of the war. The OLYMPIA's last major mission was that of bringing the body of America's "Unknown Soldier" from France to Arlington National Cemetery in 1921.

The OLYMPIA was decommissioned in Philadelphia in 1922 and berthed at the Navy Yard without care for the next twenty years, nearly becoming a contribution to the scrap metal drives of World War II. With Presidential intervention the OLYMPIA was designated as a naval relic of the Spanish-American War in 1942, but no other action was taken, not even maintenance. In 1954 the Navy tried to dispose of all historical relics except the CONSTITUTION. This threat spurred historically-minded committees to raise funds for restoration. Title to the ship was acquired by the Cruiser Olympia Association in 1957. Some repairs were made by a commercial shipyard; the work was slovenly, a large portion of the port engine disappeared, and the yard went bankrupt. A new Association was formed, and restoration is proceeding as monies are gathered; even though the ship has been declared a National Historical Monument, no government funds are provided.

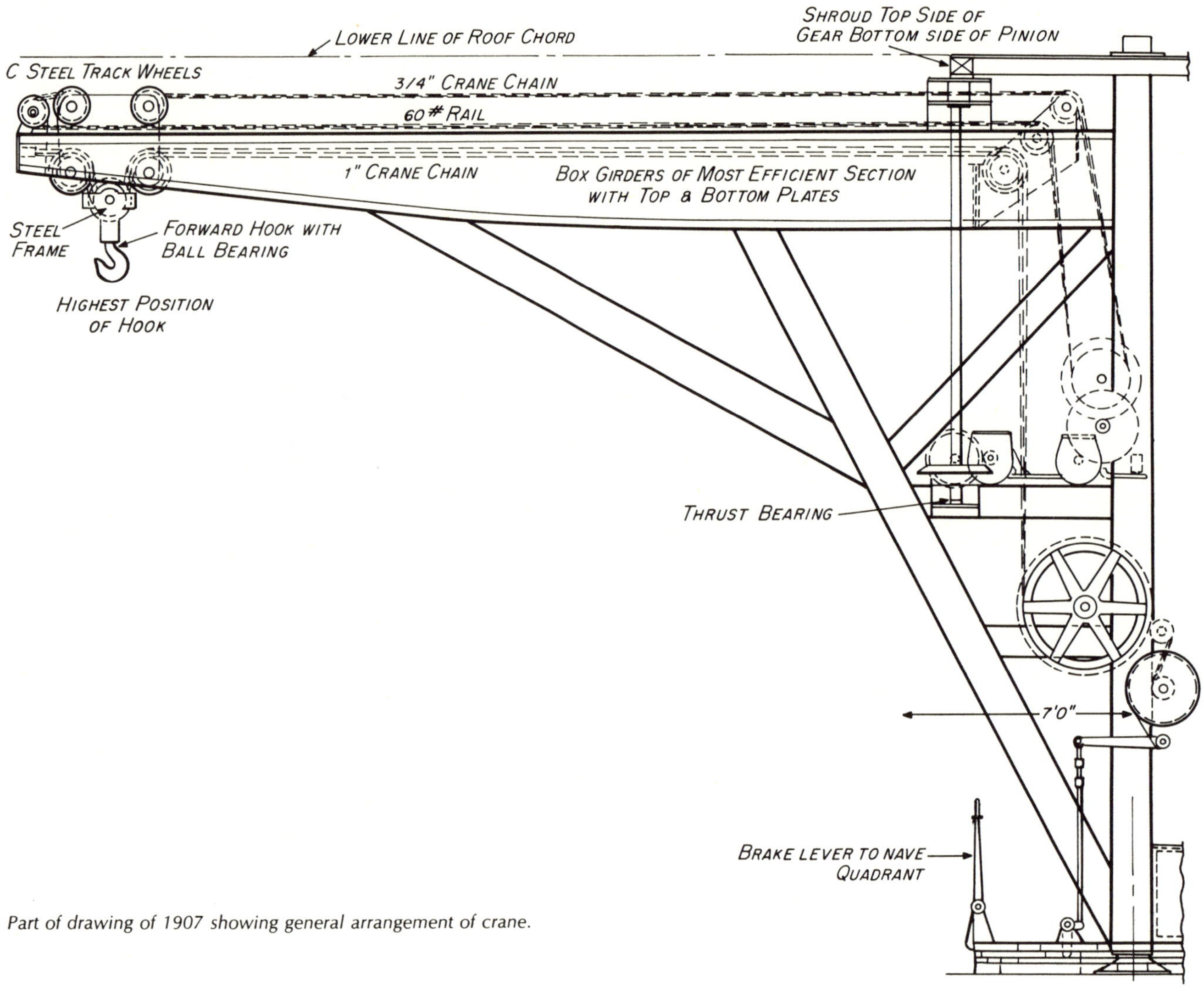

Part of drawing of 1907 showing general arrangement of crane.

Five-Ton "Pit-Cast" Jib Crane—1905

American Cast Iron Pipe Company
2930 16th Street North
Birmingham, Alabama

If the title appears enigmatic, consider that it refers to a jib crane of five-ton capacity lifting castings made in a pit: it is not that a crane of five-ton weight was cast in a pit! The castings manipulated by this crane were cast iron pipes, long, of large diameter and heavy, cast in a deep pit for convenience, safety, and the promotion of quality.

Early European and American water pipes were made of round tree trunks, unbarked, preferably elm or fir. As soon as the community could afford it, the wooden pipes whose average life-span was 14 years were replaced with more reliable cast iron pipe, as Philadelphia did, the Watering Committee laying nine-foot lengths of pipe in 1817 imported from England. The success here created a growing demand for cast iron pipe of local manufacture.

The first large scale use of cast iron occurred in the latter half of the seventeenth century, in the France of Louis XIV. It went into the mains and distribution systems of the Marly Water Works installation laid in 1664-86 that supplied the fountains and other water displays of Versailles and Marly-le-Roi, the latter establishment becoming a casualty of the Revolution in 1789. Money had been of no moment to the royal concern cast iron was very dear because of the enormous quantities of charcoal required (coke from coal came only in 1709) – so over 15 miles of cast iron piping were procured, much of it from Belgium. The meter-long sections of pipe had flanged joints fitted with leather gaskets, the lead-caulked spigot-and-socket joint coming about a century later. Diameters ranged from 1.75 to 20 in. (4.5 to 51 cm) with wall thicknesses around 1.5 in. (3.8 cm). Some of this pipe is in service today; the cast iron is in fine shape, but the wrought iron bolts have often rusted away.

Designated
March 31, 1977

Prior to 1850 cast iron pipe had been made in horizontal "green sand" molds over dry sand cores in lengths up to 9 ft (2.7 m). American industry began to offer pipe from 1819 onward, and a foundry devoted exclusively to the making of cast iron pipe was built in Millville, New Jersey in 1834.

After 1852 the vertical dry sand or pit-cast system – a vertical casting method imported from England – gave greater strength, uniformity and accuracy than the earlier green sand technique and supplanted it. The molds, up to 12 ft (3.6 m) long, were stood on end in a pit, spigot down. Impurities would rise to the top of the mold, where provision had been made for cropping the anticipated sullage. Also, should the mold open inadvertently – there was a substantial hydrostatic head due to the fluid's specific gravity of 7.2 – the consequences would not be disastrous.

The American Cast Iron Pipe Company – ACIPCO – was incorporated and chartered in 1905, with original plant facilities for the production of pit-cast pipe being constructed 1905-06. Six jib cranes about 28 ft (8.5 m) high with a 35 ft (10.6 m) jib, built by either Alliance Machine Company of Alliance, Ohio, or Cleveland Crane and Car Company of Cleveland, Ohio, were installed to preside over the pits.

A novel feature in 1906 was the use of electric motors instead of the slower and cumbersome manual or hydraulic cylinder operation. Each crane had three motors to care for hoisting, booming and slewing separately. In the beginning, the brakes for each function were mechanical, but they yielded to air brakes that were finally replaced by electrical units.

The cranes operated back-to-back, each commanding a 25 ft (7.5 m) deep pit. Upended pipe molds, lined with sand and fitted with black-washed cores to form annuli for the pipe, stood in the pits. After pouring and solidification of the metal, the mold was opened, the pipe extracted and removed from the pit – due to the office of the crane.

The crane itself is of structural steel, the generously braced box-girder jib being about 36 in. (0.92 m) deep. All loads are carried by chain (no cable), and booming (the in-out motion) takes place without a change in elevation of the hook and load during the 25 ft (7.5 m) horizontal traverse.

The Jib Crane as it appears today in general service. The foreground on which the photographer stood is the filled-in, 25-foot-deep "pit" from which the vertically "pit cast" pipe were retrieved by the crane.

By the mid-1920 s the pit-cast method had given way to one that is now computer-controlled, namely, centrifugal casting. In this, horizontal molds made of iron and rotating about their axes are filled with molten iron; only a small core for the interior of the socket is needed. The resulting pipe (spinning provides 17 g for consolidation) is of denser, more uniform and of stronger metal than before to give a far better product, being 25 percent lighter and much smoother than sand-cast pipe. Automated, the production rate is many times faster than that of the pit-cast method.

In consequence of the change to centrifugal casting, the pits have been filled and all but one of the cranes removed. This veteran of the class of '06 is still active as a general purpose crane, and goes about its business with the cheerful vigor of the old.

Aerial view of 90-acre lake fill with station and railroad facilities.

State Line Generating Unit No. 1—1929

Hammond, Indiana

Designated
April 19, 1977

For a quarter of a century – from 1929 to 1954 – Unit No. 1 Turbine Generator of the State Line Station (on the Illinois-Indiana border) was the world's largest electrical production unit with its rating of 208 mW. When placed in service, it produced 15 percent of the power of the entire Chicago metropolitan district.

The man responsible for the marvel of its day was London-born Samuel Insull (1859-1938), sometime private secretary to Thomas A. Edison, who for our purposes may be defined as a "public utilities executive known for his contributions to good utilities operations and his mistakes in utilities financing." Insull, as President of Commonwealth Edison Company, advocated the practice of supplying many communities from a central station. Even in 1919 his systems covered most of Illinois and extended into neighboring states. It was in this spirit that plans for the million-kilowatt station to be known as "State Line" were announced in 1926 as reasonable preparation for the future power needs of the industrial area identified by Chicago. In 1903 Insull had risked his reputation by installing a 5000 kW steam turbine-generator (then the most powerful in the world) in Chicago's Fisk Street station instead of following accepted practice and choosing a steam-engine-driven generator. This 5000 kW machine has been preserved and is a Landmark, having been moved back to the makers, General Electric Company, Schenectady.

The Plant is located on 90 ac (36 ha) of fill on the Lake Michigan shore, a location providing adequate water, and one readily supplied with coal by either rail or water. Sargent and Lundy were the consulting engineers, while the architects were Graham, Anderson, Probst and White. General Electric furnished the 208 mW turbine generator that was

State Line's Generating Unit No. 1 from 1929 is a General Electric 208 000-kW, three-unit turbine generator set consisting of one 76 000-kW high-pressure unit, and two 62 000-kW low-pressure units each with 4000-kW house generator.

steamed by six Babcock and Wilcox boilers furnishing 450 000 lb (205 000 kg) of steam per hour at 650 psi (4500 kPa) and 730 F (387 C) on the burning of 1-1/2 tons (1-1/5 t) of coal per minute. Imposing values for 1929, later units would have "boilers" rated at 1 350 000 lb (612 250 kg) per hour of steam at 2100 psi (14 500 kPa).

On opening, Unit I consisted of one high-pressure turbine of 76 mW and two low-pressure turbines of 62 mW each, the latter also driving 4 mW auxiliary generators to give the triple cross-compound unit its total capacity of 208 mW when turning 1800 rpm. Any of the three turbo-generators could be taken out of service to leave the other two in continuing operation. Ten years later – in 1939 – a 150 mW single-shaft, two-winding turbogenerator neighbor would be the world's largest hydrogen-cooled machine. Dethroned as champion in December 1953 by a 213 mW turbine-generator in Ohio, Unit No. 1 accepted age gracefully, and operated at 171 mW in 1977.

The Pratt Institute engine room circa 1900.

Pratt Institute Power Generating Plant—1887/1900

Off Willoughby Avenue
between Classon Avenue and Hall Street
Brooklyn, New York

Charles Pratt (1830-1891), an industrialist who grasped the possibilities of the petroleum market to become affiliated with Standard Oil early in his career, made a fortune and retired in 1874 to devote himself to philanthropic enterprises. His greatest work was the founding of the Pratt Institute in Brooklyn in 1886, embodying his ideas of a thorough and practical education. The physical Institute was to be thoroughly modern, with steam heat and both gas and electric lighting, and have a seven-story elevator in the main building, one of three constructed originally. These structures still stand, although considerably altered by now. Standard mill construction ensured a usable commercial property should the school fail!

Of the early period, it may be said that the Institute was an immediate success, the initial enrollment of 12 or 15 having grown to 3000 by the time of Mr. Pratt's death in 1891. The earliest mechanical equipment is not of our concern, although we may note that it included a 40 hp Harris-Corliss engine. The plant was coal-fired for the first two years, converted to oil in 1888, and back to coal in 1896.

Two new buildings, added in the late 90 s, increased the electrical load and forced an overhaul of the plant. The original equipment was replaced by the three Landmark engines and generators that have been in regular service ever since their installation in 1900. This facility from the turn of the century is the oldest single-cylinder reciprocating steam generating plant in the Northeast, and is a rare and perhaps last survivor of a once-common installation.

Designated
April 20, 1977

The horizontal, single-cylinder engine, often called mill engine, was the workhorse of the later 1800 s and functioned well into the 20th century. Sturdy and uncomplicated, occupying but little space (when compared to the traditional beam engine), these engines found unlimited application, being for their time as ubiquitous as the electric motor is today. In the early 1900 s it was the custom of many establishments to generate their own electricity – it was direct current then, and central stations had a limited range. It was here that the simple horizontal single-cylinder engine driving a dynamo at a leisurely 270 rpm or so was supreme, not only in factories, but in schools as well as streetcar substations out toward the end of the line.

The three Landmark engines were furnished by the Ames Iron Works of Oswego (New York) during the summmer of 1900. They are direct-connected to General Electric 75 kW generators turning at about 275 rpm. These three identical units replaced an odd lot of three machines that had driven their loads – dynamos and line shafts – by belts.

These Ames engines are of the then-esteemed single-cylinder horizontal type with 14 in. (355 mm) bore and 12 in. (304 mm) stroke. Originally designed for steam at 100 psi (689 kPa), they are now operating on 120 psi (827 kPa) steam. The electrical output is 120 volts dc, and presently supplies about one-third of the original buildings. Speed control is derived from flywheel-mounted inertia governors that vary the valve travel and hence the steam-admission opening. As supplied, the engines had balanced slide valves. Somewhere in the 1920 s piston valves were fitted via a Baker conversion kit; the piston-valve housing was simply bolted to the original valve-chest face. A full description of the engine model, published in the October 12, 1893 issue of the *American Machinist,* serves now as reference/maintenance manual, for no other drawings of parts, etc., are to be found at Pratt. As remarked, the three very durable Ames engines have been in regular service since 1900. An Edison Company report of 1929 found the engine/generator units still economical in operation, but did comment that the almost 30-year-old units were "nearing the end of their useful life."

A growing campus and enriched curricula created the need for more local power, and with 1908 a three-stage Curtis turbine driving a 150 kW generator at 2000 rpm changed the tone of the powerhouse. This unit was taken out of service in 1949, a quite good record, too. In the 1920 s three-phase, 220 volt power was brought in to supplement the direct-current supply that was in part obtained from the Edison Company.

Photographs from Scientific American *of October 6, 1888 giving some idea of Pratt Institute. Note the 40-hp Harris-Corliss engine in the foreground of top right engraving.*

THE PRATT INSTITUTE. THE SMITH'S SHOP.

THE ENGINE ROOM.

THE PRATT INSTITUTE BROOKLYN N. Y.--THE TRADES SCHOOL.

After the end of WW II commercial direct current was being phased out, but with the acquisition of war-surplus diesel-generator sets, aid was at hand and the steam plant no longer needs to run during the summer when there is no need for heating steam.

Original cars are still operating on the Duquesne Incline of 5-foot gage and 800-foot length that gives a rise of 400 feet with a grade of 58.5 percent (30.5 degrees).

The Monongahela and Duquesne Inclines—1870 and 1873

Grandview Avenue Stations
Pittsburgh, Pennsylvania

Designated
May 11, 1977

Pittsburgh's industries arrayed themselves along the banks of the Allegheny and Monongahela Rivers to leave little land for housing in the triangle defined by the rivers that had also left high hills, precipitous bluffs and rolling plateaus. The large European, strongly German work force found living space on the hilltops where land values were not enhanced by the difficulty of access. In particular, there was Coal Hill (now dignified to Mount Washington, and pierced by the Fort Pitt Tunnels) rising some 400 ft on the western bank of the Monongahela that was attractive residential and farm land even if hard to reach. The Germans had memories of the old-country "Steilbahn," called Incline here and funicular elsewhere, in which cars carrying passengers and freight were moved along a double-tracked slope by a connecting cable, the up- and down-cars more or less balancing each other.

By the 1850 s plans for building one or more inclines were in hand by land developers, the earliest charter dating from 1852. The first incline is recorded for 1854; it handled coal from a mine down to a railroad line that still exists on the river bank. Despite the desirability of inclines, further progress was slow, what with right-of-way disputes, financial problems, and the Civil War.

Inclines flourished after the 1870 s, a total of 17, mainly for passengers, being built from the valley of one of the three rivers to a hilltop, either within the city or an outlying borough. Each was designed for its purpose and location. An incline, more properly an Inclined Plane, consists of two parallel tracks, one for the "down" car, the other for the "up" car. The cars are cabled together with the intermediary of a winding drum; in addition to the hauling cable, there is a second or safety cable. The cars have an odd but practical look about them, for the underbody is built to the slope to carry a horizontal floor for passengers or freight. One incline for light passenger service had a 42 in. (1060 mm) gage, while another was a broad 10 ft gage (3040 mm). For general passenger transport and light freight the usual gage was 5 ft (1520 mm).

The first passengers-only incline was the Monongahela of 1870, the still-existing "Old Mon" as it is affectionately termed. A German mechanical engineer, John Endres, who had settled in Cincinnati in 1866, was called upon to engineer and construct the incline that went into service in May, 1870. It appears that this was the first passenger incline in America. Cincinnati, like Pittsburgh a town with hill-top communities, quickly emulated Pittsburgh's example and had its first incline in operation two years later. Endres' assistant on the Monongahela project was another European, Samuel Diescher, who went on to design, not only most of Pittsburgh's inclines to come, but also inclines in other cities, ranging even to South America. Interestingly enough, Diescher would assist G. W. G. Ferris in the design and construction of the Pittsburgh-built, 1000-passenger 250 ft (76 m) diameter Ferris wheel erected in Chicago for the World's Columbian Exposition in 1893.

The Monongahela passenger line was paralleled in 1883 by the Monongahela Freight Incline that also carried passengers and functioned until 1935. The year 1883 saw the Penn or 17th-Street Incline of 10 ft (3.04 m) gage in operation, handling 20 ton coal cars until 1953. It was followed by the Mt. Oliver incline, 1872-1951. The third incline, Duquesne Heights of 1873, is the other survivor of the past, sharing Landmark honors with "Old Mon." The pertinent facts relating to the inclines remaining in everyday operation summarize as follows:

	Monongahela	*Duquesne*
Length, ft (meters)	640 (195)	793 (241)
Elevation, ft (meters)	370 (112)	400 (122)
Grade, degrees	38	30
Grade, percent	71	57
Gage, ft (millimeters)	5 (1524)	5 (1524)
Type of car	3 compartments	1 compartment
Passenger capacity	25	18-25
Opened	May 23, 1870	May 20, 1877
Rebuilt (iron structures replaced wood)	1882	1888
Repowered (electric motors replaced steam engines)	1935	1932
Speed, mph ((km/hr)	6 (9.6)	4 (6.4)
Ridership (1975)	460,000	575,000
(1891)	1,000,000	

Some data for the original power are at hand for the Duquesne Incline. There was a double engine of 14 in. (350 mm) bore with 24 in. (610 mm) stroke; a horsepower figure is lacking but a 75 hp (56 kW) electric motor took over in 1932. The engine shaft carried a 30 in. (762 mm) pinion engaging a 12 ft (3660 mm) gear on the shaft of the winding-drum handling 1-1/2 in. (48 mm) steel cable. The 12 ft (3660 mm) bull-gear (it is

The original 12-ft diameter bull-gear of the winding drum of the Duquesne Incline has 132 wooden gear teeth set into slots, to be replaced as wear against the iron motor pinion dictates. Both pulling cables of 1-1/8-in. diameter wind onto the grooved periphery of the 12-ft diameter, almost 4-ft wide, drum. The 75-hp motor replaced the original steam engine in 1932.

still at work) has 125 inserted wood teeth of about 11 in. (280 mm) face width and a thickness of 1.8 in. (46 mm) or so: these rock-maple teeth have a casual profile, being bandsawed to shape and replaced as necessary.

Of Pittsburgh's once 17 inclined planes only two, the Old Mon and Duquesne Heights, remain in everyday operation. Both have been owned by the Port Authority of Allegheny County since 1964, with the Duquesne Incline leased to the *Society for the Preservation of the Duquesne Heights* Incline for an annual fee of $1.00. This *Society* was formed in the early 1960s when declining revenues, other means of transportation and decay combined to threaten the closing of the Incline that was under private ownership. A group of concerned citizens, mostly residents of Mt. Washington, came to the fore with plans for salvation, establishing a working agreement whereby the neighborhood would launch a fund-raising drive and the then private owners would consent to the repairing and reopening of the Incline if sufficient money was forthcoming. The *Society* raised the necessary funds, volunteers from the community cleaned, overhauled, refurbished and rehabilitated the Incline, returning it to service after a shutdown of but a few months. This Incline was recognized by the Pittsburgh History and Landmarks Foundation in 1969, and is included in the National Register of Historic Places.

The 77-foot Great Falls of the Passaic River provided the power for the development of Paterson as "The Cradle of American Industry."

The Great Falls Raceway and Power System—1794

Paterson, New Jersey

This, the third joint ASME-ASCE National Historic Engineering Landmark, is, like an earlier venture (Mount Washington Cog Railway) unrivaled, for it marks a planned community of manufacturing plants that were to secure young America's industrial independence from Europe. It was in 1779 that Alexander Hamilton, first secretary of the Treasury, projected a Federal industrial city, necessarily sited by a source of water power.

The idea took form in 1971 with the chartering of a stock company, the *Society for Establishing Usefull (sic) Manufacturers (SUM).* A six-mile square was provided within which the Society might reign. The location was at the sharp bend of the Passaic River at Paterson dominated by the Great Falls, of 77 ft (25 m) height and 280 ft (92 m) width. A dam above the Falls would provide water for the Great Falls Raceway and Power System.

Championed by Alexander Hamilton, Pierre Charles L'Enfant, the planner of Washington, drew up a set of plans that were judged too elaborate and expensive for the time. These plans, modified by Peter Colt, superintendent of the *SUM* operations in 1793, defined the first raceway that operated from 1794 to 1799, drawing water from a wooden dam to run a mill.

Financial leadership difficulties as well as operating experience caused the *SUM* to become a power developer and real estate firm in 1800. The original raceway was extended to provide service for more mills, and new lots and water rights were sold as more manufacturers came into the area. In 1820, and again in 1838, substantial additions were made to the system, as new raceways, and a masonry dam downstream of the old wooden barrier came into being.

Designated
May 20, 1977

By 1876 the development was impressive, for the *SUM* estimated that of the potential 2100 hp (1565 kW) of the raceways, 1836 hp (1370 kW) were in use. Industry of about every kind had been attracted to Paterson. Of the many and varied enterprises, a few may be mentioned. Samuel Colt's revolver, an 1835 patent, found no early acceptance, his company failing in 1842. (Success would come later, after the Mexican War 1846-48, when Colt was already in Connecticut.) There were five locomotive manufacturers in town, building nearly 80 percent of American locomotives at one period. Both of the famous Civil War engines, the Union raider "General" and the successful Confederate pursuer "Texas," were Paterson-built by Rogers Locomotive Works. The sequel, it may be remembered, was the execution of some members of the crew from the "General" as spies. By 1870, half of the silk manufactured in this country came from Paterson, called "Silk City." More recently – 1920 – Wright Aeronautical Corporation, the engine manufacturer, located in Paterson.

The *SUM* kept pace with the times, supplying its customers with electric power derived from a hydroelectric generating plant built during 1912–14 that had a steamplant backup for the low-water stage. Decommissioned in 1969, the plant will be called to life as part of the general restoration program. With 1945, the *SUM* ceased operation, selling its property, assets, charter rights, hydroplant, and raceways to the City of Paterson.

In the mid-1960 s the historic portion of the country's oldest industrial city was threatened by progress in the form of a proposed six-lane highway. An active citizen's group formed itself to preserve and restore the area, some 119 acs (49 ha), stretching from the Great Falls at the river to downtown Paterson, known as the *Great Falls/SUM National Historic District.* It was designated the nation's first *National Historic District* by President Ford in June 1976.

Among other things, the District holds some 49 mills, many of which produced the silk for which Paterson was so well known. A paper-making mill dates from 1803, a cotton mill from 1812. Colt's gun factory of 1835 survives also. The mills will not resume their original tasks, but will instead function as craft studios, specialty shops, art galleries, studios, and the like, in addition to supplying other commercial and office space. Restaurants will overlook the waters, and there will be sheltered paths along the old raceways.

A major effort is being placed on the Erecting Shop of the Rogers Locomotive Works. This shop, dating from 1871, replaced earlier structures on land leased from the *SUM* in the 1830s. The building is a fine example of the early use of cast iron columns and brick bearing walls supporting heavy timber beams. The Rogers Shop will serve as a combined Cultural Arts Center and Museum for the District and house some civil offices.

It may be mentioned that the upstream West Side Park has a plaza mounting an early submarine, the "Fenian Ram," built in 1881 by John Holland, a local schoolmaster. Holland's submarine of 1878 that sank in the same year to lie in the river for 50 years, is in a local museum.

Reproduction of the Vulcan Street power plant of 1882, opened in 1932 on the occasion of the plant's 50th anniversary. The claim of being the "World's First Hydro-Electric Central Station" has since been revised to being only "the first Edison hydro-electric central station to serve a system of private and commercial customers in North America."

Vulcan Street Power Plant—1882

807 South Oneida Street
Appleton, Wisconsin

Designated
September 15, 1977

This hydroelectric central station of 1882 is the first Landmark receiving triple recognition, the American Society of Civil Engineers and the Institute of Electrical and Electronics Engineers having joined the American Society of Mechanical Engineers to designate the Vulcan Street Plant a National Historic Civil, Mechanical, and Electrical Engineering Landmark. The present Plant was made for the Golden Jubilee Anniversary in 1932; it is not on the original site.

In the beginning, electricity was used only for illumination, being the upstart that was competing with gas; indeed, in some areas, electrical development was hindered by the gas interests. While the arc lamp had its place, it was far too bright and otherwise impractical for home and small-space lighting purposes. Incandescent lamps had been tried as early as 1820, but all had had severe shortcomings. After much travail, Thomas A. Edison found that split bamboo when carbonized made a satisfactory high resistance filament on the 110 volts he had chosen to reduce transmission losses that lower voltages would have imposed.

The earliest application of the Edison lamp system was aboard an arctic expedition steamer (the JEANNETTE) that had been fitted with generator and electric lights in 1879 that worked well until the vessel was crushed in the ice two years later. Some 150 Edison-installed electric light plants were at work in ships, residences, mills, stores, offices and other places by 1882. In New York, Edison's Pearl Street Station, had, with steam engines totaling 900 hp (670 kW) the capacity to operate 7200 lamps at 110 volts when it went into service on September 4, 1882. England had put a 3000-lamp station on the line a bit earlier, in January of that year.

September 1882 was a momentous month, for one of the very earliest hydroelectric plants, the forerunner of the Vulcan Street Plant – began operation on September 30. Back in 1882, and even in 1932, local enthusiasm identified the installation as the "world's first hydro-electric central station." This statement has been revised: The plant is the first Edison hydroelectric central station to serve a system of private and commercial customers in North America.

The genesis of this plant is related in a newspaper, "The Appleton Post" noting in its edition of July 27, 1882, under a paragraph headed "Light Transaction," that "Mr. H. J. Rogers is a man who deals in heavy things, such as paper mills, pulp mills and ponderous powers. [Now] . . . he has gone in for light transactions. He has purchased the exclusive right for illuminating the towns of the Fox River Valley with the Edison electric lamp. The near future will develope [sic] this investment into another fortune to be added to his millions, and it serves him right." Rogers, who was president of the Appleton Paper and Pulp Company as well the Appleton Gas Light Company, had moved quickly in getting his "exclusive right" from the Western Edison Electric Light Company of Chicago, an outfit that had been incorporated only two months earlier (May 25), for the licensing of Edison plants in Illinois, Wisconsin, and Iowa.

Appleton, located on a bluff above the Fox River, had made good use of the river's 33 ft (10 m) drop within city limits, water power being applied to mills of various kinds. Mr. Rogers had the first electric generator, an Edison Type "K" dynamo of 12.5 kW or 250-lamp capacity installed in the beater room of his paper-and-pulp mill. This mill, together with the Vulcan Paper Mill and Mr. Roger's own pretentious residence on the height above the river, had been wired for lights. It was September 30 when the lights went on and the paper would report that the illumination was "bright as day."

With the generator driven by a wheel that was part of a mill system with fluctuating load, the lights went from dim to bright, and sometimes burn-out at $1.60 per bulb. Accordingly, the generator was relocated at the mill and given its own wheel. The second generator, also an Edison Type "K," fared even better, being given its own building on Vulcan Street in November. A regional inventor, Mr. Elmer of Berlin, Wisconsin, had a patent for a waterwheel – a turbine runner – that seems to have had local popularity.

The Edison Type "K" generator, gear- and belt-driven from a vertical-shaft Elmer waterwheel.

The Vulcan Street plant had an Elmer wheel of perhaps 42 in. (1060 mm) diameter that turned at 75 rpm under a 10 ft (3 m) head of water brought in by a 28 in. (710 mm) penstock made of 3 in. (76 mm) timber. Bevel gearing and belt-drive to the generator increased the speed several-fold. The output of 12.5 kW was capable of lighting 250 sixteen-candle-power lamps, each equivalent to 50 watts.

The total installation, having been hurried so, was quite rude. There were neither voltmeter nor ammeter, let alone customer meters: the charge was simply so much per lamp per month. Voltage regulation – speed control – was handled manually, the operator judging the state of things by the brightness of the lamps. There was no lightning protection, nor main fuses. Smaller fuses, electricity's safety valves, were rather casual, being mounted on wooden bases; wood was also used for sockets and other components. Bare copper wire formed the distribution lines, although lighter gage stuff for house wiring was insulated with cotton wrap. With the poor insulation, small disturbances meant short-circuits that had all hands out looking for the trouble while service was suspended. Wires were supported by wooden cleats, tape being used for extra insulation when wire was poked through holes in partitions.

By December of 1882 three or more homes, five or six mills and a blast furnace were lighted by Edison bulbs. The Waverly House, regarded as one of the finest hotels of the middle west, became even more popular on being wired for lighting in 1883. Two larger generators of 85 kW each were acquired in 1886 and placed in a new central plant to which the original dynamos were also moved, Appleton changing over to a three-wire distribution sytem – one of the world's earliest – at this time. The new plant had other advanced features, as regulating devices and fuses. Customer meters came with 1888. In 1890 twenty-four-hour service (instead of dusk to dawn) was begun, and a few customers installed motors. With 1886 there was even an electrified street railway that continued in use until 1930, yielding then to buses. Since the cars could be controlled from only the front, turntables were needed at the end of the line. The 60 kW generator for the cars was water-powered. There was no lighting protection, so service was shut down during storms.

The Slater Mill Historic Site on the west bank of the Blackstone River.
Left to right: Wilkinson Mill, 1810; Sylvanus Brown House, 1758; and Slater Mill, 1793.

The Wilkinson Mill—1811

Pawtucket, Rhode Island

Designated
October 12, 1977

Pawtucket, adjacent to Providence at the head of deep-water navigation of the Pawtucket River, and served by the water-power of the 50 foot falls of the Blackstone River, became one of the first American textile centers by the end of the 18th century. Earlier, in 1670, the attractive site had been settled by Joseph Jenks, a skilled iron-worker making domestic utensils and tools. The village became the center for distinguished and inventive artisans who would provide the mechanical skills for the construction (and later, design) of textile machinery that would spread through the United States.

Among those who came to Pawtucket was Oziel Wilkinson, a clever Quaker blacksmith who moved his family from nearby and very rural Smithfield in the 1780 s. Dealing first in nails, he expanded to open a water-powered smithy and ironware shop. Anchors for the local shipyard, and large handmade iron screws for the presses of the paper, cloth and oil trades were among the heavy items of the Wilkinson enterprise.

Manufacturing America was in a sorry state until long after the Revolution, for it was strongly affected by a Parliamentary Act of 1774, an Act that contributed no small amount to the War of Independence. Enacted shortly after Arkwright's textile-machinery inventions had inaugurated the factory system in Britain, the Act forbade, under heavy penalties, the exportation of machines used in the manufacture of cotton, silk, wool, and linen goods. There was a prohibition, too, of the migration of experienced mechanics who had worked in English factories. On the other side, several American legislative bodies offered bounties and rewards to skilled workmen, as a hundred-pound award made to a man for a carding machine.

Chief among those responding to the American lures was Samuel Slater who had risen from management trainee to overseer in a Derbyshire cotton-spinning plant filled with Arkwright's inventions. He arrived in New York in 1789 to find immediate employment. Unimpressed with both the machinery and the motive power, he soon followed another lead to Moses Brown of Providence whose firm was struggling to develop a cotton-spinning mill. It should be noted that while the weaving of plain textile goods had been speeded up and made less dear by Kay's invention of the flying shuttle in 1733, handspinning of thread would continue for another generation or so. Quicker weaving increased the pressure on spinners to produce thread more efficiently than spinning by wheel, but this challenge was not met until the advents of Arkwright's water-frame (1769), and Crompton's mule (1779). In America, just as it had been in Britain, the vital and missing element was spinning machinery. It was here that Slater made his great contribution, for from memory and experience he designed a set of machinery "all on the Arkwright principle" on which the first cotton thread in the United States was spun by water power in the now successful factory of Almy and Brown located in an old fulling mill. Sylvanus Brown, woodworker and an expert mechanic whose house is also to be seen, had executed the machinery under bond of secrecy, the iron-work being furnished by David Wilkinson, son of Oziel. Slater went on to build, in 1793, what is now known as the Old Slater Mill, restored and maintained by the *Old Slater Mill Association.*

Slater had, in effect, provided "instant industrialization," but this accomplishment was possible in part because of skilled mechanics who could turn ideas into hardware and make contributions of their own. These mechanics virtually initiated the machine tool industry of the United States as they anticipated the demands of the burgeoning textile operations and other fields.

Slater's machinery was so effective that the mills outpaced the supply of raw cotton and operations were headed for shutdown but for Eli Whitney's fortuitous invention of 1793, the cotton gin. The slave who had been able to clean – de-seed – only a pound of cotton in a day could now turn out fifty or more. Domestic cotton which had been grown casually was now carefully cultivated on large plantations, with the consequences of an increase in slavery and a one-crop system that terminated in the Civil War. In 1792 America produced less than 2 000 000 lb (907 t) of cotton; by 1800, production stood at

35 000 000 lb (15 900 t), and in 1845, America was responsible for seven-eighths of the world's output.

David Wilkinson was a talented mechanic who would be responsible for a great machine-tool tradition. He was also a steamboat pioneer of sorts, having built a steam engine for the propulsion of a boat fitted with far from promising goose-foot paddles that ran after a fashion in 1793. At any rate, the engine experience would be drawn upon later for powering mills.

In 1794-95 David's father Oziel had constructed a water-powered rolling and slitting mill to the south of the Slater mill to produce rolled plate and nail rod. By 1796 David had added a water-powered machine for cutting large screws. Aside from its function, this machine and its derivative, the industrial lathe, would be unique on two counts. Firstly, the machines formed from heavy castings had a rigidity and stability impossible with the older wooden-framed machinery. Secondly, the cutting tool was clamped to an iron carriage — we call it a tool slide or slide rest today — that traversed the length of the workpiece by locking a split nut onto a leadscrew. We may note again the two features commonplace today but quite novel at the end of the eighteenth century — massive castings providing a rigid frame, and a tool no longer hand-held but instead clamped to a carriage given positive motion by a lead screw. Similar developments were going on in France and Britain at this time, but all the evidence indicates independent development on the part of Wilkinson. With some justice, Wilkinson has been credited with being the founder of the American machine tool industry.

But to return to the screw-cutter, in which a lead screw and blank workpiece were in parallel and connected by gears having a one-to-one ratio. There is casual and vague mention of an 18-20 ft (5.5 to 6 m) dimension which is more likely the overall length of the stout base or machine frame than the length of the workpiece or potential screw. The cutting tool that engaged the blank was clamped to a carriage rolling on three wheels for statically determinate support on what must have been a somewhat uneven guideway of the bed casting. This carriage was akin to what in smaller and wheelless format would be called a slide rest, as in a conventional lathe.

Having no change gears (at least, none are mentioned), the machine could only cut screws (and also nuts) that were of the same lead as the master screw. From what little is known, the machine was hardly a precision tool, but it did a job and was a machine of industrial size and capacity. A patent was granted in 1798, but even here the data are scanty because of the Patent Office fire of 1836.

Wilkinson's next and greatest tool, an offshoot of the screw-cutting machine, was an industrial lathe, large for its purpose but smaller than its ancestor. This machine, so common today, represented a great step forward in both the history of American machine tools and the development of American industry. A heavy-duty, general-purpose tool, it was a lathe rather than a single-minded screw-cutter. Its outstanding features were the screw-driven slide rest and tool, the former three wheels having been dispensed with in the smaller machine, and, of course, the rigidity imparted by castings.

This lathe was never patented, and much information is lacking since it exists in neither artifact nor picture. For example, it is not even known whether it had change-gears. One likes to think that it did, but this is speculation. Nevertheless, it was recognized in this country as an eminent tool, and its basic principle, the tool-carrying slide-rest under the firm control of a lead screw, was widely copied. And well it might be, for it meant that the pitiful handheld tool operation of metal-turning was a thing of the past: where metal had been removed with the manually held, long-handled "hook tool" in the manner of wood-turning, it was now done with a mechanically applied tool-point. It was a really incredible advance.

These powered, powerful and productive machines quickly found their ways into shops, factories, and federal arsenals. The merit of the tool is testified to by a $10,000 award given Wilkinson in 1846 by the Congress for the invention of the "sliding lathe" that was "a striking illustration of the force of American genius."

It was in 1811 that the Wilkinson Mill, a stone building designed for cotton spinning, was erected just across from the rolling and slitting mill. For power it had a waterwheel and was backed by a steam engine made by David Wilkinson. Of the steam engine nothing is known except that it existed.

The waterwheel is more interesting. We may note that while a "traditional" waterwheel was outside its building, and had a diameter many times its width, such wheels

Failed cast iron shafting used as lintels over bricked-in tailrace openings. The lefthand lintel shows the diminution of section near the broken-off journal of the shaft.

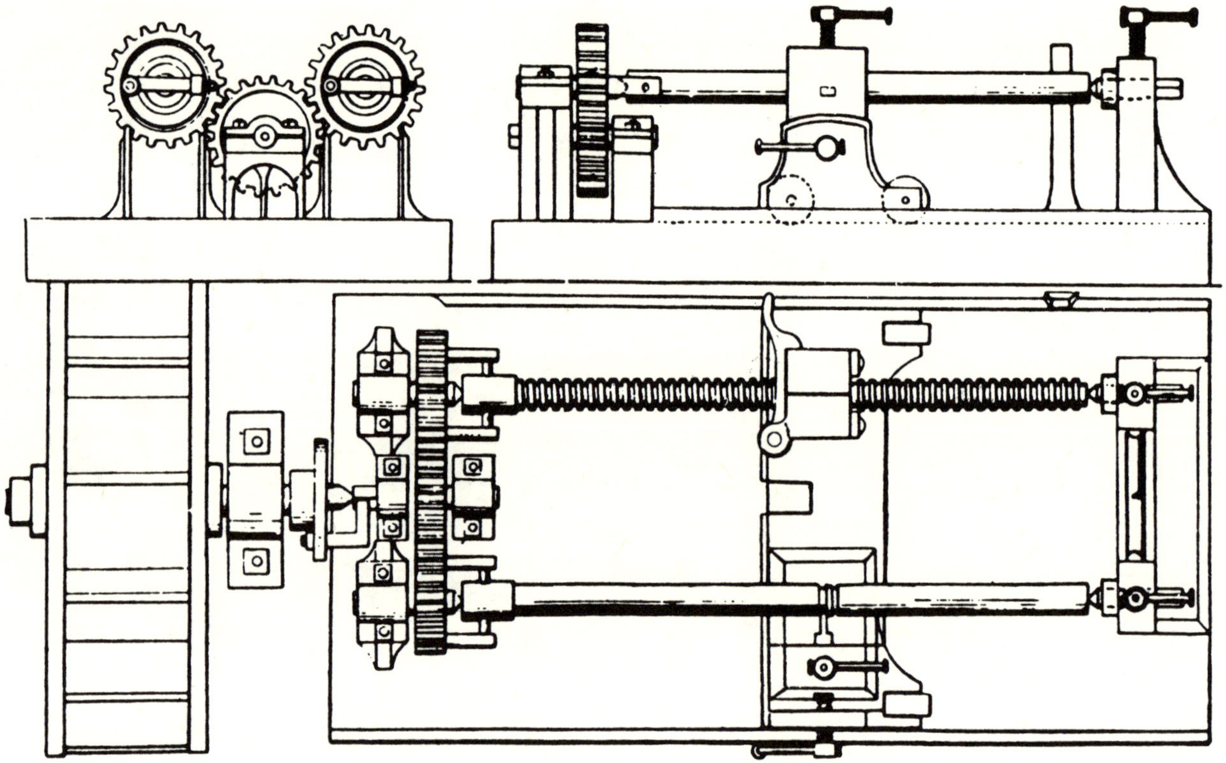

David Wilkinson's historic screw-cutting lathe patented in 1798. The overall length of the base may have been 18–20 feet. The tool carriage, akin to a modern slide rest, rolled on three wheels for steady support.

generally drove only a single machine and were of low power. With textile mills, however, the wheel was brought into the mill building and given a central location that the multitude of little machines might be driven. A different wheel geometry allowed more power — the wheels were "long," in the present case 9 ft (2700 mm), while the diameter was a stout 12 ft (3600 mm). (An English mill of a little earlier date had a wheel of 12 ft (3600 mm) diameter that was 48 ft (14 600 mm) long and built around a hollow wooden shaft of 8 to 9 ft (2400-2700 mm) diameter that was hooped like a cask!) The primary power transmission was an almost incredible complex of cast iron gears and cast iron shafting of square or diamond cross section except where journaled for bearing support. Line shafting on the several storys was driven by bevel gears from vertical shafts geared to the waterwheel, the gearing providing the necessary increase in shaft speed. The individual machines were belted to the line shafts.

David Wilkinson, largely responsible for the operation of the mill, presided over a machine shop on the first floor. This shop dealt not only with the local maintenance problems but also built new machinery for local use and commercial sale. From this latter activity Wilkinson achieved a reputation as a master builder of textile mill equipment that was deployed not only in New England but into other areas too, as reaching down to Georgia and across to Pennsylvania. The power loom that eventually dominated weaving in New England for much of the nineteenth century was a descendant of the first machine that David built from inventor William Gilmour's plans in 1817.

Cotton ceased to be king in the Wilkinson mill in 1819, for it became the site of one of the first worsted manufactories in the United States when an English emigrant, William Wood, and David, set up a spinning frame of 72 spindles that employed seven workers. The census-taker of 1820 spoke very highly of the cloths that included silk.

In later life (1846) Wilkinson recalled a machine he had made about 1820 for the purpose of grinding spindles for jennies. The description is superficial but tantalizing, and although there is neither drawing nor artifact, it might be that Wilkinson had built the first centerless grinder. Today the centerless grinding of cylindrical work provides a high production rate, high finish and great accuracy at low cost.

The mill had power problems, causing one to wonder what the standby steam engine may have been like. Somewhere between 1826 and 1835 the original 9 ft-long (2700 mm) waterwheel was replaced with one of 13 ft (3900 mm) length but of the old 12 ft (3600 mm) diameter, and a new tailrace was added. The lintels over the second race openings were sections of failed cast iron shafting, and are still to be seen. Despite its estimated 10 hp the new wheel proved to be inadequate; after its removal, power was brought in by a long shaft from the waterwheel of the old slitting mill.

Textiles suffered a severe depression in 1829 causing David Wilkinson to lose his business and leave Pawtucket. The next owner held the Mill about thirty years, but the innovative period was gone, a variety of disparate interests centering in the old mill until 1901, when the Pawtucket Electric Lighting Company bought the building. Ultimately the Old Slater Mill Association acquired the property for restoration and adaptive use. As a first measure of success the architectural restoration has been honored by an award from the New England Regional Council of the American Institute of Architects.

Bits and pieces of the original installation, as sections of both vertical and horizontal cast iron shafting, parts of textile machinery, the location of the second wheel and a portion of its structure, framing, and the like, will serve as authentic guides to a reconstruction of the wheel and its entire drive to the mill's machinery. On completion it will be the only example of an early nineteenth-century breast wheel driving a North American textile mill.

The Wilkinson influence extended far beyond the mere construction of successful production machinery, for his shop was also the training ground for the first generation of textile machinists, a group collectively responsible for innovations in textile technology.

The Paige Compositor of 1887, a machine to set foundry type, that, because of its intricate and complicated nature, lost out to a younger contemporary, the Mergenthaler Linotype.

The Paige Compositor—1887

The Mark Twain Memorial
351 Farmington Avenue
Hartford, Connecticut

A printed page is the consequence of numerous operations, each vital in its own way. Not the least important is the assembling of the type to make-ready for printing. Hand-composition, or the picking of type by hand and arranging them in a type-form, was, in the nineteenth century, the principal hindrance to getting on with the job of printing. Type-casting was well-mechanized, but the design of machines that would set type ready for printing at a rate greater than hand-composition occupied many inventors from the early 1800 s on. Two kinds of machines were envisioned. One might be a typesetter that would draw on a ready supply of type – letters, figures, punctuation marks and the like – and put them all together, hopefully justifying the lines as well. The other machine, instead of selecting pre-cast type from a magazine, would cast its own as a preliminary to assembling and justifying. It would be this last kind of machine – the Mergenthaler Linotype – that would be the winner in the struggle that was not resolved until just before the turn into the twentieth century: it was a machine minimizing mechanical complication, not needing to sort and distribute and store type-slugs after the press run.

The Paige Compositor of 1887 preserved in *The Mark Twain Memorial* is a machine of the first kind; it was a runner-up in the contest, and had it appeared twenty years earlier, might well have been a success, at least for a time. The Paige machine has passed into American folklore with the legend that it bankrupted Twain. It is true that he was a heavy investor – "he lost a fortune" – but not even his biographers are sure of how much, although a maximum figure seems to be $300,000. Twain's ill fortune with the machine was compounded by the panic of 1893 and the failure of his publishing house: he did manage financial recovery.

Designated
December 7, 1975

This machine, carrying the name of James W. Paige, had substantial contributions from Charles R. North, E. S. Pierce and Charles E. Davis, among others. It was twenty years in development, and the present machine is the survivor of the two built before the effort was terminated in 1894. The development costs ran over two million dollars. The machine came to fruition just as the more versatile and less complicated Linotype machine matured.

A ponderous five thousand pounds is packed into a nine-foot length. The machine is the most elaborate and intricate typesetter ever produced, having over 18 000 parts. Withal, it could set only one size type (agate), but it did justify the lines to 0.005 inch. The board of 109 keys was so arranged that a skilled operator (using all fingers of both hands!) could bring down whole words, the types sliding into a raceway for assembly followed by justification of the line. Only one operator was needed instead of the three demanded by other typesetting machines. Three columns of "dead" type could be examined for damaged matter that was discarded, the rest being distributed into their separate channels after proper orientation.

There was a second machine, completed in 1894 and installed in the office of the Chicago Herald for a sixty-day trial and demonstration. It did very well, needing but one operator instead of the usual three, out-performing the other thirty-two machines in use. However, the printing industry was inclining toward the less complex typecasting compositors. The machine was purchased by the Mergenthaler Linotype Company and presented to Cornell University. It came to an unworthy end, being callously junked during a World War II scrap drive.

Detail of Reynolds-Corliss engine of circa 1917.

Reynolds-Corliss Engine and Pump—1917

Main Street Pumping Station
Jacksonville, Florida

Designated
February 23, 1976

Jacksonville's water supply improvement program of 1914-17 saw the installation of two 5,000,000 gallon-per-day pumps, each driven by a Corliss engine. These pumps were the city's sole supply until 1930 when the present electrically-driven peripheral pumping stations began to assume the load. Steam operation ceased in 1956, and the first pump of 1915 was scrapped. The second pump, installed in 1917, remains on a possible standby basis; even though declared a Landmark, it may be called to serve again in time of crisis.

This historic machine was built by Allis-Chalmers, and consists of a pump coupled to a Reynolds-Corliss engine. The engine is of particular interest as a surviving local example of the Corliss engine patented in 1849 and improved by later engine-builders, in this case by the chief engineer of Allis-Chalmers. Corliss engines are distinguished by having four semi-rotary valves per cylinder, two for steam and two for exhaust, set at right angles to the centerline of the cylinder. The valve-ports are short, and entering steam does not pass through ports previously cooled by the exhaust. An oscillating wrist-plate makes for quick-acting valves susceptible to fine tuning of the inlet valve for cutoff, and of the exhaust valve at the point of release. All in all, the economy of the Corliss was some 35 percent better than that of contemporary slide-valve engines. The design became popular in the United States and was widely copied by European engineers. Corliss' patent expired in 1873, after which "everyone" was free to use the idea and did.

The present engine carries the name of Edwin Reynolds who spent most of his life with Corliss engines. After ten years as superintendent of the Corliss Works in Providence, Rhode Island, Reynolds went to E. P. Allis in Milwaukee in the capacity of superintendent in 1877. With the formation of the Allis-Chalmers Company in 1901, he became chief engineer of that firm. Reynolds was responsible for numerous improvements, notably in the valve-gear and governor relations to create an even more distinctive engine, whence the coupling of his name with that of Corliss. A wide variety of engines for blowing, pumping, hoisting, rolling mills and electrical generators were built during his tenure. He held some 48 patents, and was President of ASME in 1902.